Study Guide

Biological Systems of Animals

Level 3 Technical Animal Management

Second Edition

The publisher gratefully acknowledges the permission of copyright holders to reproduce copyright material: Fig 1 designua©123RF.COM; Fig 2 tassel78 /Shutterstock.com; Fig 3 grayjay/Shutterstock.com; p.8 (top) alila©123RF.COM; Fig 4 Alila Medical Media/Shutterstock.com; Fig 5 Alila Medical Media/Shutterstock.com; ImagineDesign/Shutterstock.com; Fig 7 Biology Education/Shutterstock.com; p.12 udaix/Shutterstock.com; Fig 8 Alexander_P/Shutterstock.com; Fig 9 Designua/Shutterstock.com; Fig 10 Aldona Griskeviciene/Shutterstock.com; Fig 11 EreborMountain/Shutterstock.com; Fig 12 Aldona Griskeviciene/Shutterstock.com; p.18 Designua/Shutterstock.com; Fig 13 decade3d - anatomy online/Shutterstock.com; Fig 14 decade3d - anatomy online/Shutterstock.com; Fig 16 Tefi/Shutterstock.com; Fig 18 martinpel©123RF.COM; Fig 19 (left) Breck P. Kent/Shutterstock.com; Fig 19 (right) cedecea©123RF.COM; p.28 (both) Anviczo/Shutterstock.com; Fig 20a designua©123RF.COM; Fig 20 poemsuk©123RF.COM; Fig21 Nicholas Taffs/Shutterstock.com; Fig 22 gritsalak karalak/Shutterstock.com; p.33 andegro4ka©123RF.COM; Fig 23 SciePro/Shutterstock.com; Fig 24 SciePro/Shutterstock.com; Fig 25 decade3d©123RF.COM; Fig 26 Daniel Eskridge/Shutterstock.com; Fig 27 Sergey Uryadnikov/Shutterstock.com; Fig 28 VectorMine/Shutterstock.com; Fig 29 jtplatt©123RF.COM; p.43 3drenderings/Shutterstock.com; Fig 30 designua©123RF.COM; Fig 30 and p.51 Isselee©123RF.COM; Fig 32 and p.56 Eraxion©123RF.COM; Fig 33 Sakurra/Shutterstock.com; Fig 34 normaals©123RF.COM; Fig 35 VectorMine/Shutterstock.com; Fig 36 Marina Veder/Shutterstock.com; Fig 37a mik38©123RF.COM;Fig 37 and page 68 About time/Shutterstock.com; Fig 38 designua©123RF.COM.com; p.69 tigatelu©123RF.COM.com; Fig 39 designua©123RF.COM; Fig 40 normaals©123RF.COM; Fig 41 geki©123RF.COM; Fig 42 and p.84 (upper) guniita©123RF.COM ; Fig 43 Alila Medical Media/Shutterstock.com; Fig 44 sonnydaez©123RF.COM; p.75 (left) Alan Tunnicliffe/Shutterstock.com ; p.75 (right) kostya6969©123RF.COM; Fig 45 Designifty/Shutterstock.com ; Fig 46,47a alexilus©123RF.COM; Fig 47 Blamb/Shutterstock.com; Fig 49 (left) rixie©123RF.COM ; Fig 49 (right) adogslifephoto©123RF.COM; Fig 50 designua©123RF.COM; Fig 50a sciencepics/Shutterstock.com; p.84 (lower) maxcreatnz/Shutterstock.com; Fig 51 worldswildlifewonders/Shutterstock.com; Fig 53 angeluisma©123RF.COM; Fig 54 zsolt_uveges/Shutterstock.com; Fig 55 Agnieszka Bacal/Shutterstock.com ; Fig 56 John Ceulemans/Shutterstock.com; Fig 57 hakbak1979©123RF.COM; Fig 58 creativenature©123RF.COM; Fig 59 Designua/Shutterstock.com

Cover image: SciePro/Shutterstock.com

Figure 17 p.25 © M Drost (1974). Reproduced with kind permission of Dr Maarten Drost at the University of Florida.

Figure 48 p.79 and p.85 © Laurie O'Keefe, reproduced with kind permission - www.laurieokeefe.com

Figure 52 p.87 © Thomas Haslwanter. This work was made available under the Creative Commons Attribution-ShareAlike 3.0 Unported licence. Details of the licence are here: https://creativecommons.org/licenses/by-sa/3.0/deed.en
The original work can be found here: https://commons.wikimedia.org/wiki/File:LateralLine_Organ.jpg.

All other photographs and illustrations are © Eboru Publishing.

Every effort has been made to trace copyright holders and to obtain their permission for the use of copyright material. The publisher will be glad to make arrangements with any copyright holder it has not been possible to contact.

All rights reserved. No part of this publication may be reproduced, distributed, or transmitted in any form or by any means, including photocopying, recording, or other electronic or mechanical methods, without the prior written permission of the publisher, except in the case of brief quotations embodied in critical reviews and certain other noncommercial uses permitted by copyright law.

Copyright © 2023 Eboru Publishing

First edition 2021. Second edition 2023. Impression 10 9 8 7 6 5 4 3 2 1

This second edition has been prepared with help from Hannah Norgate.

ISBN 978-0-9929002-6-7

Ordering Information

Special discounts are available for class set purchases by schools, colleges and others. For details, contact the publisher at: enquiries@eboru.com

Trade orders: copies of this book are available through the normal wholesalers. For any queries please contact: orders@eboru.com

www.eboru.com

Features in this book

Topic introduction
Brief summary of what you will cover in the next section.

In this topic you will learn about:

- The components that make up the circulatory system, including blood and blood vessels.

- The components and function of the double circulation system, including the detailed structure and function of the heart.

Questions
Knowledge-check questions at the end of each section, so that you can quickly recap on what you have learned.

Questions

1 What is an erythrocyte and what is its function?

2 What is a leukocyte and what is its function?

Link
This feature highlights the connections between different topics within the unit.

Link
As blood is pumped around the body it is filtered by the kidneys, which excrete waste from the body - see section 1.4.

Did you know?
Interesting facts and background information.

Did you know?
Whilst fish respire using gills, some ancient species evolved lungs to allow them to breathe air.

Glossary and keywords
Important words that you should be able to define are shown in **bold**.

Useful words, that you may not need to know for your exam, but which might be helpful for your understanding, are shown in red.

Both sets of definitions are at the end of the book.

Contents

LO1 Understand the structure and function of biological systems in animals — 6

1.1 Structure and function of the circulatory system	6
1.2 Structure and function of the respiratory system	13
1.3 Structure and function of the reproductive system	19
1.4 Structure and function of the excretory system	29
1.5 Structure and function of the musculoskeletal system	34
Learning Outcome 1 Revision Checklist	44

LO2 Understand control mechanisms in animals — 47

2.1 Structure and function of hormonal mechanisms in the endocrine system	47
2.2 Structure and function of the lymphatic system	52
2.3 Structure, function and adaptations of the thermoregulatory system	57
Learning Outcome 2 Revision Checklist	60

LO3 Understand the neural control mechanisms in animals — 62

3.1 Gross anatomy of the brain	62
3.2 Neural control mechanisms in animals	64
Learning Outcome 3 Revision Checklist	69

LO4 Understand how animals' senses have adapted to their environment — 71

4.1 How animals' senses are adapted to their environment	71
The eye	71
The ear	76
The nose	79
The mouth	80
Touch	82

4.2 Specialised senses	86
Learning Outcome 4 Revision Checklist	91

Glossary 92

Useful words 98

Index 100

Answers to questions are available by signing up to the mailing list at www.eboru.com

LO1 Understand the structure and function of biological systems in animals

1.1 Structure and function of the circulatory system

In this topic you will learn about:

- The components that make up the circulatory system, including blood and blood vessels.
- The components and function of the double circulation system, including the detailed structure and function of the heart.
- The single circulatory system.
- The open circulatory system.

All the cells that make up an animal's body require oxygen, nutrients and water to function correctly. Cells also produce waste products that need to be removed. The circulatory system is responsible for transporting all of these substances around the body. It also has some other functions, including regulating body temperature and helping to prevent disease.

The main parts of the circulatory system are:

- blood
- the heart
- veins
- arteries
- capillaries.

Blood

Blood is the fluid that travels around the circulatory system. In vertebrates it is made up of:

- erythrocytes
- leukocytes
- platelets
- plasma.

Erythrocytes

Erythrocytes are also called red blood cells.

- They are responsible for delivering oxygen to other cells in the body. They contain haemoglobin, which carries oxygen.
- They do not contain a cell nucleus.
- They make up a large percentage of blood volume.
- They are formed in bone marrow, which are areas of soft tissue within bone cavities.
- They have a **biconcave** shape. This means both sides of the cell curve inwards. This gives the cell a greater surface area for oxygen absorption.
- They are flexible, which allows them to squeeze through small capillaries.

Leukocytes

Leukocytes are also called white blood cells.

- They are part of the immune system that protects the body against pathogens.
- Unlike erythrocytes they contain a cell nucleus.
- They make up a small percentage of blood volume.

- They are found in blood and in the lymphatic system.
- They are also formed in bone marrow.

> **Link**
> For an explanation of the lymphatic system see section 2.2

There are a number of different types of leukocytes, that perform different functions. There are five main types:

- Neutrophils can engulf and destroy pathogens such as bacteria and viruses. (This process is called phagocytosis). They are among the first responders to any foreign bodies. Dead neutrophils are a major constituent of pus.

- Eosinophils can also attack pathogens, including larger ones such as parasitic worms, but also cause allergic reactions to foreign substances.

- Basophils are the least common type of white blood cell and play a role in allergies. They contain histamine, which causes inflammation during an allergic reaction. (You may know that medication for relief of hayfever and other allergies contains antihistamines).

- Monocytes can differentiate into two different forms (macrophages and dendritic cells) but both play a part in cleaning up dead cells, phagocytosis of pathogens and identifying invaders to lymphocytes. They are much bigger than neutrophils and live for longer. They are responsible for inflammation and swelling.

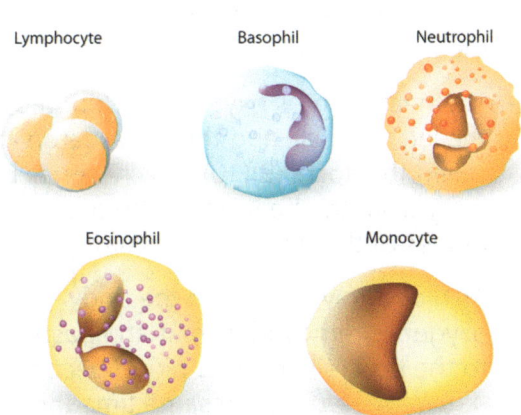

Figure 1 The five types of white blood cell

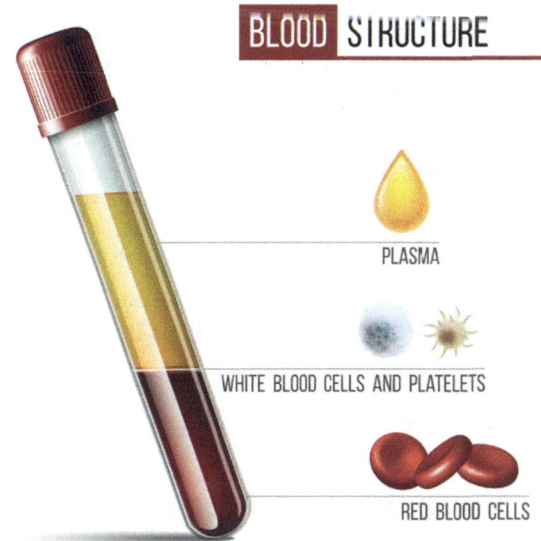

Figure 2 The components of blood

- **Lymphocytes** are part of the adaptive immune system, responsible for creating antibodies for specific foreign bodies and are discussed in more detail in section 2.2.

You may read in other sources that these five types of cells can also be arranged into further subcategories, according to their physical appearance or other characteristics.

Platelets

- **Platelets** are responsible for the formation of blood clots using fibrin. Blood clots form to slow down and stop bleeding.
- Platelets are irregular-shaped cells and do not contain a nucleus.
- They make up a small percentage of blood volume.
- They are also formed in bone marrow.

Plasma

- A liquid in which the erythrocytes, leukocytes and platelets are suspended.
- **Plasma** is yellow in colour and makes up a large percentage of blood volume.
- It transports nutrients, waste products, water, enzymes and **hormones** around the body.

How blood clots work

When a blood vessel is damaged platelets begin to bind to the wound site. A series of chemical reactions begin, which turn a substance called fibrinogen, which is soluble in blood plasma, to a substance called fibrin, which is not soluble in blood plasma.

Fibrin forms a net or web near the wound site. It acts like a scaffold for platelets to stick to, until the blood in that area becomes so thick that it clots.

Blood clots are essential to prevent infection from pathogens and to stop bleeding from minor wounds.

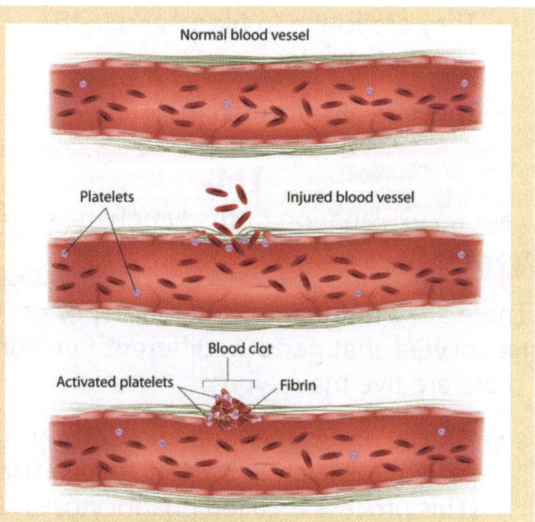

Components and function of the double circulation system

Mammals, birds, reptiles and amphibians have a **double circulation system**. This means that to complete one full circuit blood passes through the heart twice.

The heart is a muscle which is constantly contracting then relaxing to form a heartbeat. It acts like a pump which pushes blood around the body.

A simplified version of the double circulation system is shown in Figure 3.

- Deoxygenated blood is pumped once from the right side of the heart to the lungs, where it picks up oxygen.

- Oxygenated blood returns from the lungs to the left side of the heart, where it is pumped around the body.

The left side of the heart has thicker muscle so it can pump blood all around the body under high pressure.

A more detailed examination of the process is shown in Figure 4 and described below:

- When an animal breathes in, oxygen enters the lungs and from there is passed into the blood. Blood that is rich in oxygen is called oxygenated and is shown in red in the diagram.

- This oxygenated blood circulates from the lungs through the **pulmonary veins** (6) to a chamber of the heart called the **left atrium**.

- Blood travels from the left atrium through the **bicuspid valve** (7) and into the **left ventricle**. The bicuspid valve only allows blood to flow in one direction, from the left atrium to the left ventricle. The bicuspid valve is also known as the mitral valve or the left atrioventricular (AV) valve.

- The heart pumps oxygenated blood out of the left ventricle and through a large artery called the **aorta** (9)(10), which carries blood away from the heart to all the organs and tissues of the body via capillaries.

- As blood travels around the body, oxygen is removed from it and passed to the body's cells. This blood becomes deoxygenated. At the same time the cells pass waste products and carbon dioxide into the blood, so they can be removed from the body.

Figure 3 Double circulation system

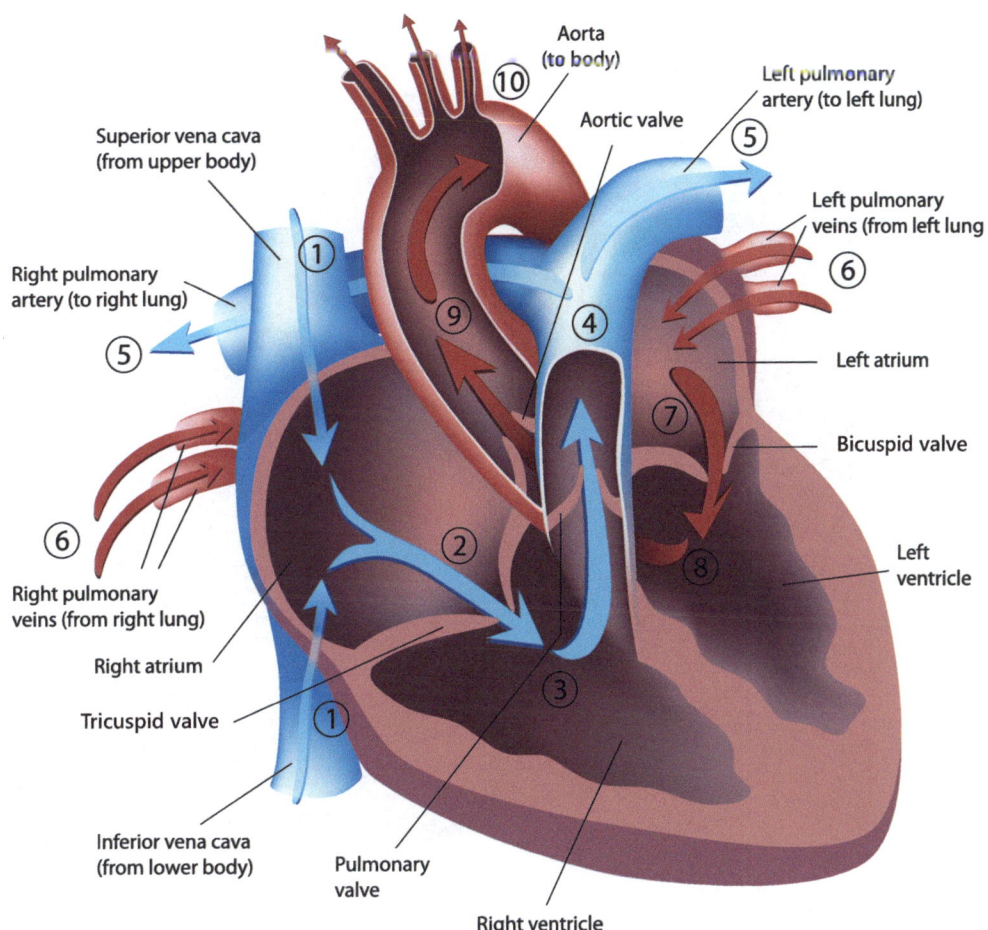

Figure 4 Blood flow through the heart

- The deoxygenated blood, shown in blue, flows from two large veins called the **vena cava** (1) and into **right atrium** of the heart (2). (Note: in Figure 3 'superior vena cava' and 'inferior vena cava' refer to the 'upper' and 'lower' vena cava.)

- From there it flows through the **tricuspid valve** and into the **right ventricle** (3). The tricuspid valve is also known as the right atrioventricular (AV) valve, and it prevents blood flowing back from the right ventricle into the right atrium.

- From there, the deoxygenated blood is pumped by the heart through the **pulmonary artery** and back to the lungs (4), (5), where it can receive more oxygen and begin the cycle again.

The **chordae tendinae** (not shown in Figure 4) are cords of tissue (sometimes called the heart strings) that are connected to the bicuspid and tricuspid valves to ensure that the valves work correctly.

The action of pumping blood around the body is caused by each atrium and ventricle contracting and relaxing in a coordinated manner. These contractions are caused by electrical signals and transmitted by special structures as shown in Figure 5:

- The **sino-atrial node,** located in the right atrium, is the source of the electrical signals. It generates an electrical signal which travels across the right and left atrium and causes them both to contract, pushing blood from each atrium into the ventricles (stages 2 and 7 of Figure 4).

- The **atrioventricular node** is located between the right atrium and right ventricle and passes on the electrical signal into the ventricles. However, it delays the signal slightly so that each atrium has a

> **Link**
> As blood is pumped around the body it is filtered by the kidneys, which excrete waste from the body - see section 1.4.

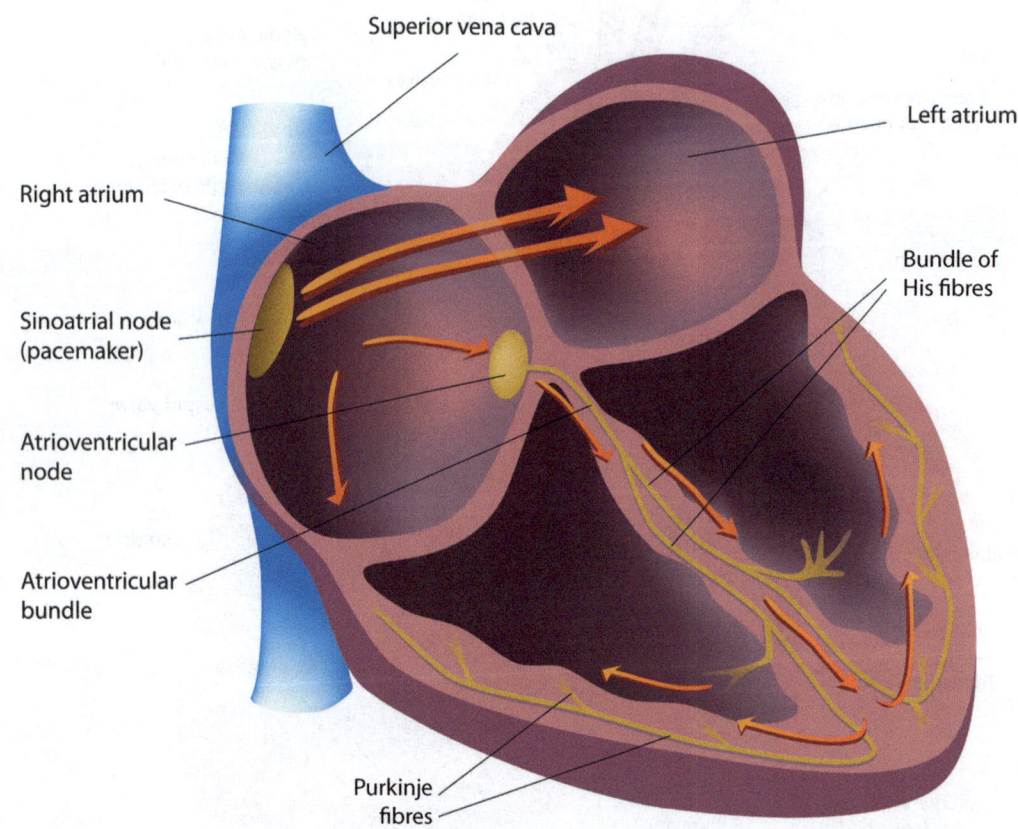

Figure 5 How the heart beats

chance to contract, and fill the ventricles, before the ventricles themselves contract.

- Connected to the atrioventricular node are **Bundles of His fibres** which transmit electrical signals down the walls that separate the two ventricles. (Notice the capital 'H' in His – this is because they were discovered by a Swiss scientist called Wilhelm His.)

- At the bottom of the ventricles are **Purkinje fibres** which further transmit the electrical signals to the bottom of the ventricles and cause them to contract. At this point the ventricles are full with blood, so the contraction forces blood out of the heart (stages 4 and 9 of Figure 4.) These fibres are named after a Czech scientist who discovered them, called Jan Purkinje.

The heart is made of a special kind of muscle called cardiac muscle. Cardiac muscle never gets tired, which is why the heart can keep beating all day and night. For more on different types of muscle see section 1.5.

Relative structure and function of blood vessels

Blood is transported around the body in blood vessels. There are three types of blood vessels: arteries, veins and capillaries, as shown in Figure 6.

Arteries

- **Arteries** take blood away from the heart.
- Almost all arteries carry oxygenated blood (though one exception, labelled as (5) in Figure 4, is the pulmonary artery which takes deoxygenated blood from the heart to the lungs).
- Arteries carry blood at high pressure.
- Arteries have thick, muscular, elastic walls – which means that the channel in which blood flows (called the lumen) is narrow.

Veins

- **Veins** carry blood towards the heart.

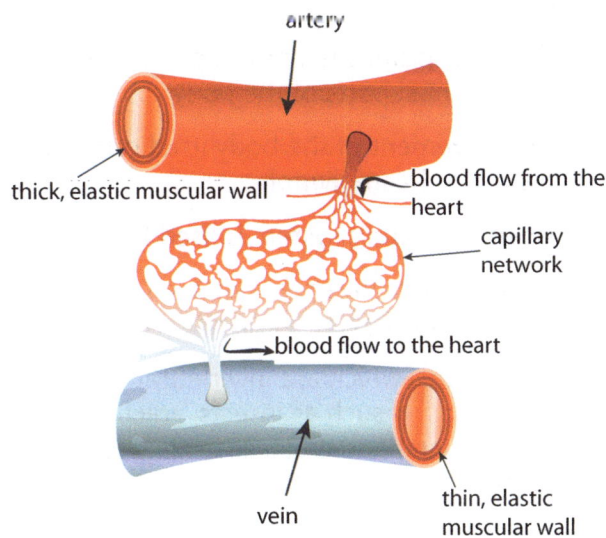

Figure 6 Arteries, veins and capillaries

- Almost all veins carry deoxygenated blood (though one exception, labelled as (6) in Figure 4, is the pulmonary vein which takes oxygenated blood from the lungs to the heart).
- Veins have thin, less muscular, elastic walls - which means that the lumen is wide.
- Veins often contain valves to keep blood moving in the right direction, particularly when blood is moving in the opposite direction to gravity (e.g. blood flowing up the legs back towards the heart).

Capillaries

- **Capillaries** are a network of tiny blood vessels that connect arteries and veins together.
- It is the capillaries that allow oxygen and waste products to be transferred to and from blood, in order to reach the surrounding tissue and organs.
- This happens because the capillary walls are only one cell thick and molecules, such as oxygen and carbon dioxide, can easily pass through these walls.

Closed circulatory systems

A **closed circulatory system** means that the blood is entirely enclosed within blood vessels (i.e. arteries, veins and capillaries). Mammals, birds, reptiles, amphibians, fish and some invertebrates have a closed circulatory system.

Single circulatory system

Whilst the **double circulation system** is present in mammals and birds, fish use another closed circulatory system called the **single circulatory system**. In the single circulatory system:

- blood only passes through the heart once to complete a full circuit of the body
- the heart only has two chambers – one atrium and one ventricle
- the heart only receives deoxygenated blood
- blood receives oxygen (from the fish's gills) after being pumped from the heart

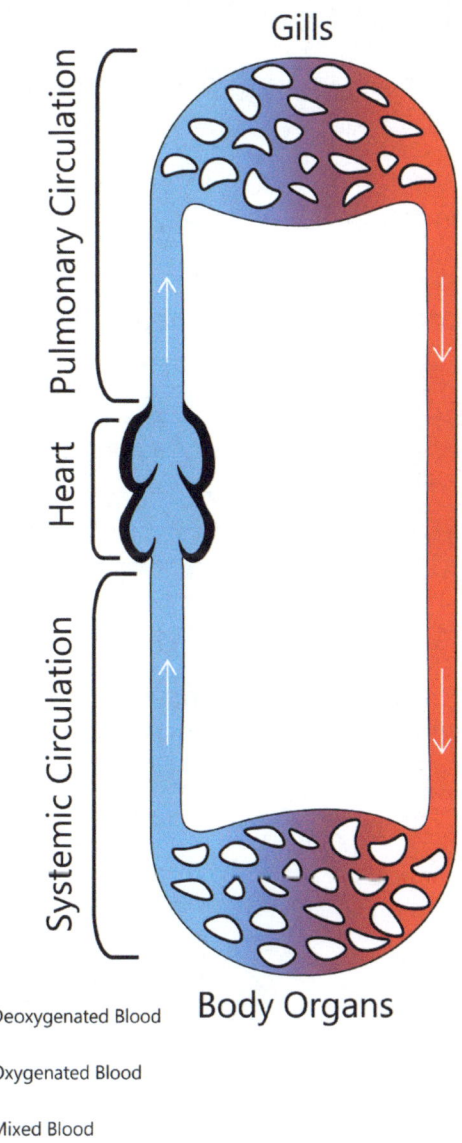

Figure 7 The single circulatory system

1.1 Structure and function of the circulatory system

- blood pressure is lower than in the double circulatory system.

Open circulatory system

In an **open circulatory system**:

- the circulating fluid is called **haemolymph**, which is the invertebrate equivalent of blood
- haemolymph transports nutrients to the cells in the body
- haemolymph is not contained within arteries, veins and capillaries - instead the whole body cavity is filled with haemolymph so the organs and tissues are bathed in it
- the heart pumps haemolymph so that it can reach all parts of the body
- the movement of the body itself can help distribute haemolymph.

The open circulatory system is used by:

- molluscs (e.g. invertebrates such as mussels, snails)
- arthropods, which are invertebrates with exoskeletons such as insects, arachnids (e.g. spiders) and crustaceans (e.g. crabs, lobsters).

Questions

1. What is an erythrocyte and what is its function? (2)

2. What is a leukocyte and what is its function? (2)

3. State the names of the following components of the heart: (6)

   a _____
   b _____
   c _____
   d _____
   e _____
   f _____

4. List three differences between an artery and a vein. (3)

5. Discuss the differences between the single circulatory and double circulatory systems. (4)

1.2 Structure and function of the respiratory system

In this topic you will learn about:

- The structure and function of the mammalian respiratory system.
- Comparative adaptations to the respiratory system in fish, amphibians, birds and invertebrates.

All the cells within an animal's body require oxygen to function. They also produce carbon dioxide as a waste product. The respiratory system is the way in which animals' bodies obtain oxygen and get rid of carbon dioxide.

Mammalian respiratory system

In mammals, the respiratory system is made up of the following structures (Figure 8):

- **Nasal chambers:** The interior of the nose, this is where air enters the body. The nasal chambers contain mucus, which warms and moistens the incoming air. They are also lined with **cilia**, which are tiny hair-like projections that trap pathogens and stop them from entering the lungs.
- **Pharynx:** Also known as the throat, the pharynx funnels air towards the larynx. It is made up of smooth muscle which allows food to be swallowed.
- **Larynx:** Also known as the voice box, this is responsible for the production of sounds. The larynx moves air towards the trachea and contains the *epiglottis*, which is a flap that stops food from entering the trachea.
- **Trachea:** This tube carries air into the lower respiratory system. It is made of C-shaped cartilage which keeps the airways open. It

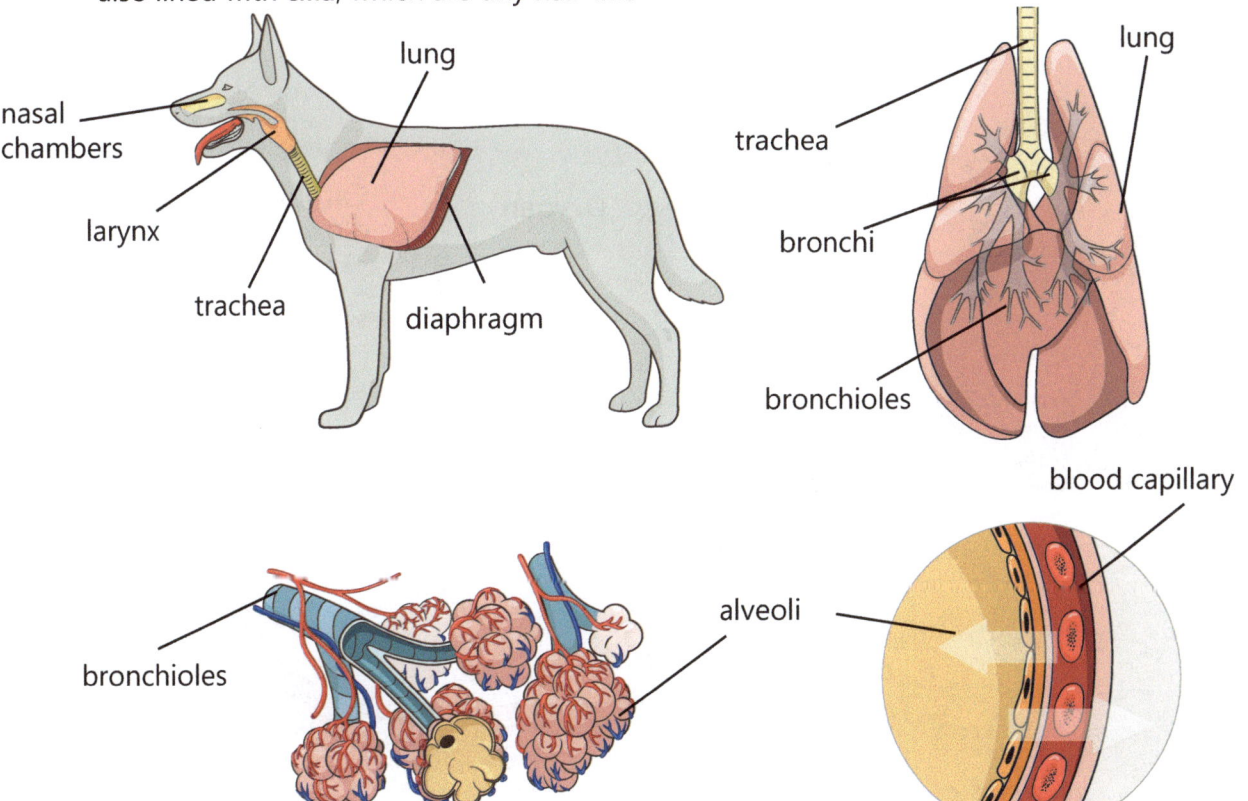

Figure 8 The respiratory system of a dog

13

also has smooth muscle which is used for coughing.

- **Bronchi:** The two main passageways that carry air into the lungs, one for each lung. Bronchi are made up of rings of cartilage and smooth muscle.

- **Bronchioles:** Branches that form from the bronchi and which spread air throughout the whole lung. They are made of smooth muscle. Air travels through the bronchioles to the alveoli.

- **Alveoli:** At the end of the bronchioles are alveoli. They are in contact with blood capillaries from the circulatory system to allow oxygen and carbon dioxide to pass in and out of blood. This is known as **gas exchange**. Alveoli are moist and only one cell thick, to allow gas exchange to take place. They also have a very large surface area, which increases the rate of gas exchange.

- **Diaphragm:** A muscle which is responsible for breathing. It is made of skeletal muscle (see section 1.5). It works with the intercostal muscles to allow ventilation - see next section.

Gas exchange and ventilation

Ventilation is a more formal term for breathing and takes place as follows:

- Contraction of the diaphragm and the intercostal muscles, which are the muscles between the ribs, enlarges the chest cavity. The diaphragm moves down whilst the intercostal muscles move the rib cage up and out.

- The increased volume of the chest cavity decreases the pressure, which draws air into the lungs via the respiratory system. This is called inhalation.

- Air reaches the alveoli deep in the lungs.

- The surface of the alveoli is surrounded by blood capillaries which are carrying deoxygenated blood.

- Oxygen molecules pass from the alveoli into the blood capillaries At the same time carbon dioxide molecules pass from the blood capillaries into the alveoli.

- The alveoli are moist because oxygen and carbon dioxide must dissolve in a liquid before they can enter a cell.

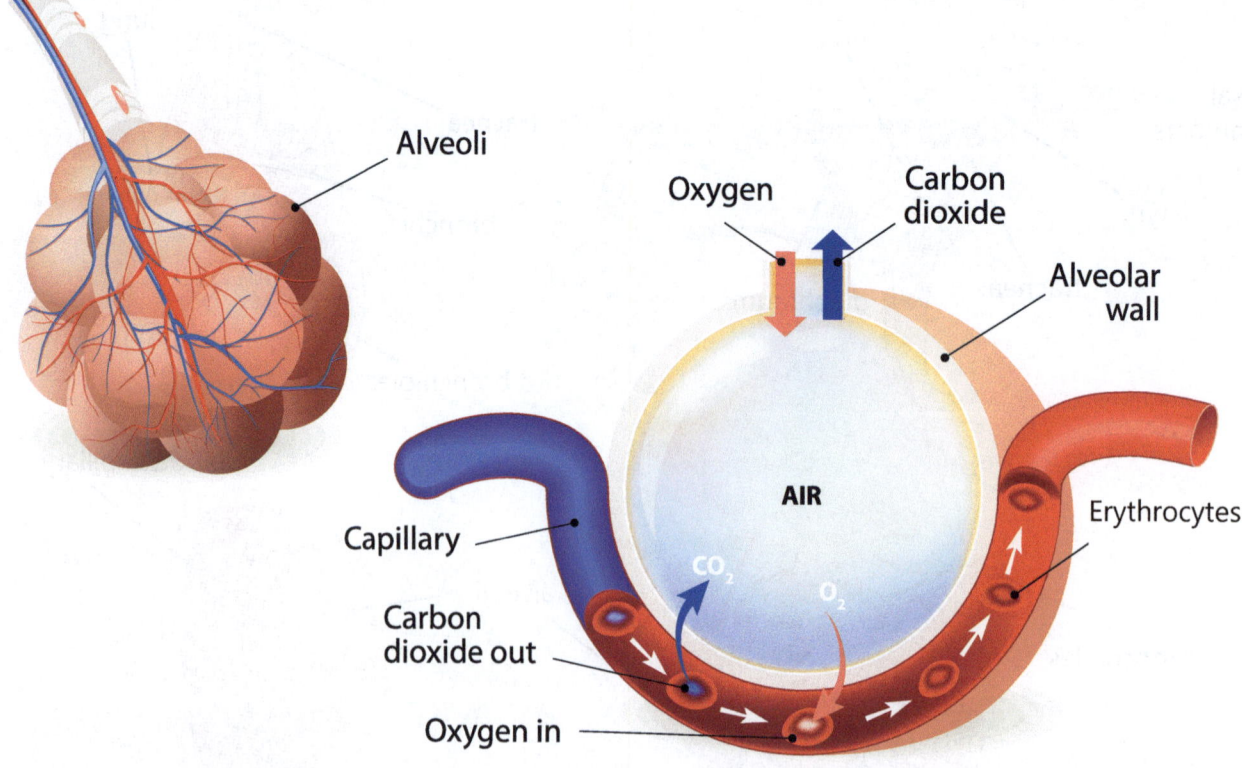

Figure 9 Gas exchange in alveoli

- Oxygenated blood travels through the capillaries, to the pulmonary vein and on to the heart.
- Relaxation of the diaphragm and intercostal muscles decreases the volume of the chest cavity, which increases the pressure. This increased pressure forces oxygen-depleted air and carbon dioxide out of the lungs, to breathed out of the nose. This is called exhalation.
- Note: most animals do not use their mouth for ventilation.

There are millions of alveoli in each lung. This, together with their grape-like structure, provides a very large surface area for gas exchange. This allows enough oxygen to be absorbed so that each of the animal's cells can function properly.

Adaptations in mammals

Whilst all mammals have the features described above, the special circumstances of some mammals have led to adaptations to the respiratory system. For instance:

- Bats are the only flying mammals. They have larger lungs relative to their body size than other mammals. Their alveoli are also smaller and more densely packed with capillaries than other mammals, which increases the surface area. This allows for rapid gas exchange. These adaptations allow a bat to extract more oxygen than other mammals This is needed because of the extra oxygen demand of flying. (Birds have adapted differently - see next section.)
- Cetaceans (whales, dolphins and porpoises) breathe through a blowhole on the top of their head. Blowholes are similar to nostrils and connect to the trachea.
- Diving mammals, such as whales and seals, have a unique set of conditions to deal with, and have the following adaptations:
    * **Their lung is fully collapsible.** Pressure increases rapidly with water depth, and this causes the lung to collapse. The structure of their lung has adapted to allow this to happen without any negative consequences. One benefit of a collapsed lung is that it reduces buoyancy – if the lung was filled with air then the animal would find it hard to dive (a bit like trying to dive in a swimming pool while wearing inflatable armbands).
    * **They use blood as a store of oxygen.** Terrestrial mammals store oxygen in their lung – this is why they take a deep breath before diving underwater. However, diving mammals use their blood to store oxygen. They breathe out before diving to allow their lung to collapse. They can store oxygen in their blood because they have relatively more blood than other mammals, and a higher percentage of oxygen-carrying erythrocytes.

Adaptations in non-mammals

The mammalian respiratory system, as described above, evolved over a long period of time to fully adapt to their natural environment and unique needs. There are some differences in the respiratory systems of non-mammals, because they evolved to adapt to different circumstances.

Amphibians

In their juvenile phase amphibians live entirely in water and respire using gills, like fish. When they become adults they lose their gills and develop lungs. However, amphibian lungs are more primitive than mammals' so they also need to breathe through their skin to supply all of their oxygen needs. They can breathe through their skin in water or air.

Amphibian skin respiration is aided by mucus on their skin which allows allows oxygen and carbon dioxide to pass through it, and into the capillaries beneath the skin. Amphibians must always keep their skin damp by living near water or in very damp conditions.

These two respiratory mechanisms (lungs and skin) ensure they always have enough oxygen.

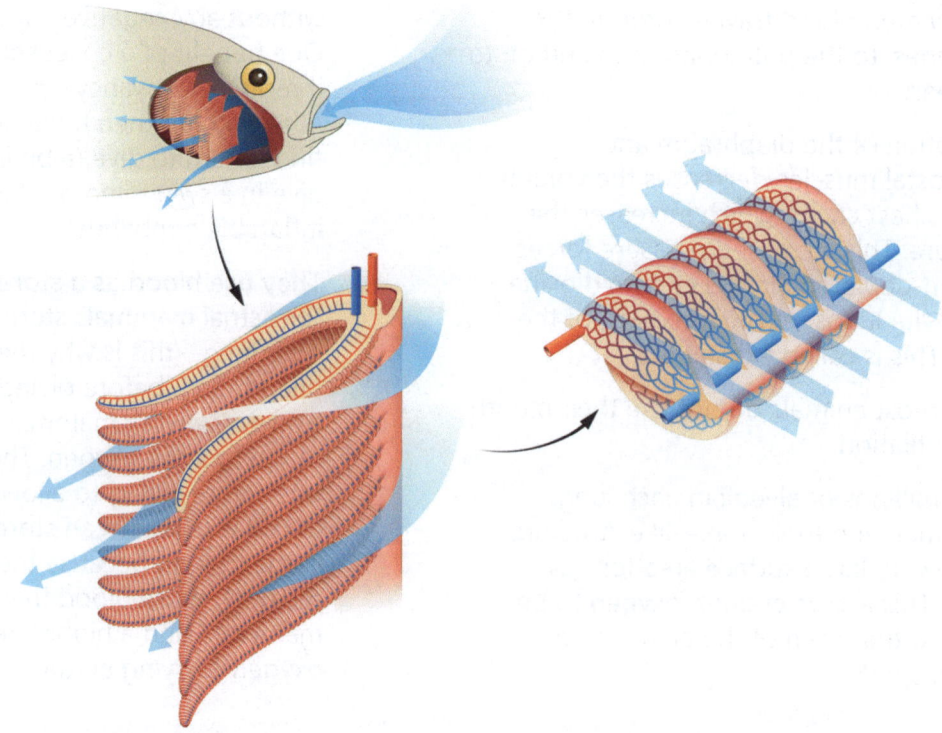

Figure 10 Fish respiration using gills. Water continually flows in one direction through the gills. Deoxygenated blood (in blue) absorbs oxygen via special cells in the gills and becomes oxygenated (in red).

Fish

In fish, gas exchange takes place at the **gills on each side of their head**. Fish use gill arches to keep the gills open and use their mouth to draw water through the gills.

There is far less oxygen in water than there is in air, so gills have to be more efficient than lungs. Fish use counter-current gas exchange, where water in the gills flows one way whilst blood flows in the opposite direction. It means that, unlike in lungs, oxygen is continually entering the bloodstream. Gills are also efficient because they have a very large surface area.

Sharks and rays have to swim constantly to move water over their gills.

Note that fish breathe oxygen dissolved in water, not the oxygen that makes up water.

Did you know?

Whilst fish respire using gills, some ancient species evolved lungs to allow them to breathe air. Such fish are the ancestors of amphibians. The modern lungfish is the only living air-breathing fish.

Birds

Birds have a much higher oxygen demand than mammals due to the exertion of flying. This requires them to extract more oxygen from air than mammals. As a result, there are a few differences from the mammalian respiratory system, most notably:

- they do not have a diaphragm
- instead, they have additional organs called **air sacs** within cavities in their body, and which act like bellows to draw air in and push air out
- their lungs do not inflate and deflate
- air only moves through the lungs in one direction (unidirectional) - this allows gas exchange to continually take place during inhalation and exhalation, which is more efficient for extracting oxygen
- it takes two inhalation-exhalation cycles for air to make its way through the system and back out again.

With reference to Figure 11, the inhalation and exhalation cycles are as follows:

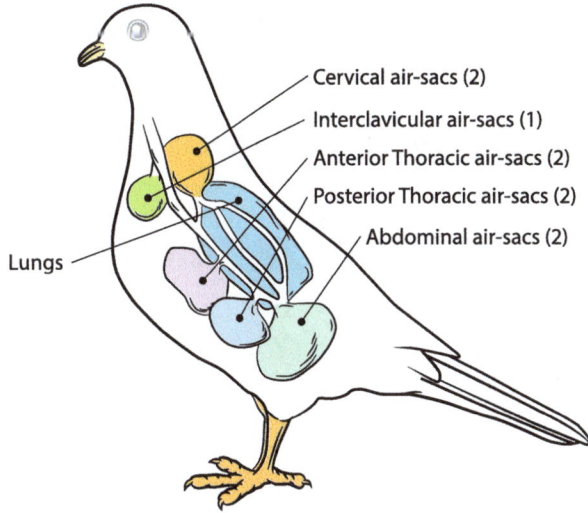

Figure 11 The respiratory system of birds

Inhalation:

- The air sacs expand and draw in air in from the mouth, which passes into the posterior and abdominal air sacs.

- At the same time, air that was already in the lungs passes into the anterior air sacs.

Exhalation:

- When the air sacs contract, air is forced from the posterior and abdominal air sacs into the lungs.

- Gas exchange occurs in the lungs – but instead of alveoli birds have **parabronchi**, a network of small tubes along which gas exchange takes place. Unlike alveoli, which are 'cul-de-sacs', the parabronchi are open at both ends, which allows for the unidirectional flow of air in the lungs.

- At the same time, air in the anterior air sac is forced out into the trachea and through the mouth.

The expansion and contraction of the air sacs is caused by the movement of muscles in the chest.

As well as allowing for greater exertion, this more efficient respiratory system allows birds to fly at high altitudes, where there is less oxygen available.

Invertebrates

Invertebrates have quite different respiratory systems.

Invertebrates such as grasshoppers do not have lungs, or even a circulatory system. Instead they have a system of tubes called **trachea** which transport air directly to cells around the body. The trachea are connected to **spiracles**, which are external openings on the surface of the invertebrate's body. Some invertebrates can control the flow of air through the trachea.

Some invertebrates (including some spiders) have a respiratory organ called a **book lung**. It consists of a series of plate-like structures arranged closely together like the pages of a book. Air surrounds the plates, which are filled with haemolyph (similar to blood) and gas exchange occurs across the surface. Book lungs are located within the abdomen, and air enters through a spiracle.

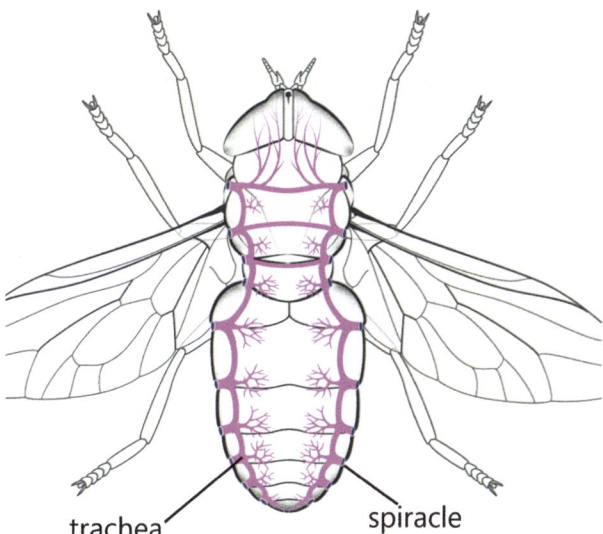

Figure 12 The respiratory system of invertebrates

Questions

1. Explain the purpose of the respiratory system. (2)

2. Discuss the differences between the mammalian and bird respiratory systems. (4)

3. What are the structures labelled A and B on the following diagram? (2)

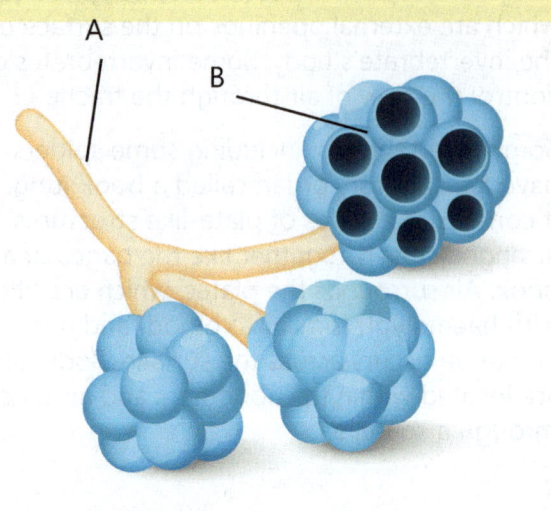

4. Describe the function of structure B. (3)

5. Describe the differences between the invertebrate and the mammalian respiratory systems. (4)

1.3 Structure and function of the reproductive system

In this topic you will learn about:

- The location, structure and function of the male and female reproductive systems.
- Comparative adaptations of the reproductive system in different species.
- The oestrus cycle.
- The stages of sexual reproduction in placental mammals.
- Adaptations to sexual reproduction, including oviparous, ovoviviparous, egg-laying mammals and marsupials.

All organisms need to reproduce. There are two types of reproduction:

1. Asexual reproduction only requires one parent. All the offspring are genetically identical to each other and to the parent. They are clones of the parent.
2. Sexual reproduction requires two parents. When the gametes (sperm and egg) fuse they create offspring that are genetically different to both parents. Different offspring are also genetically different to each other.

The rest of this section will focus on sexual reproduction in mammals, although many different organisms use sexual reproduction.

Male reproductive system

- **Penis**: It is made up of erectile tissue that fills with blood during copulation. This allows for penetration and ejaculation in the female's vagina.
- **Prepuce** (not shown): Protective sheath that prevents the penis getting damaged.
- **Bulbus glandus** (dogs): A bulb of erectile tissue at the base of a dog's penis that swells up during mating, immediately after penetration. It locks or ties the two mating

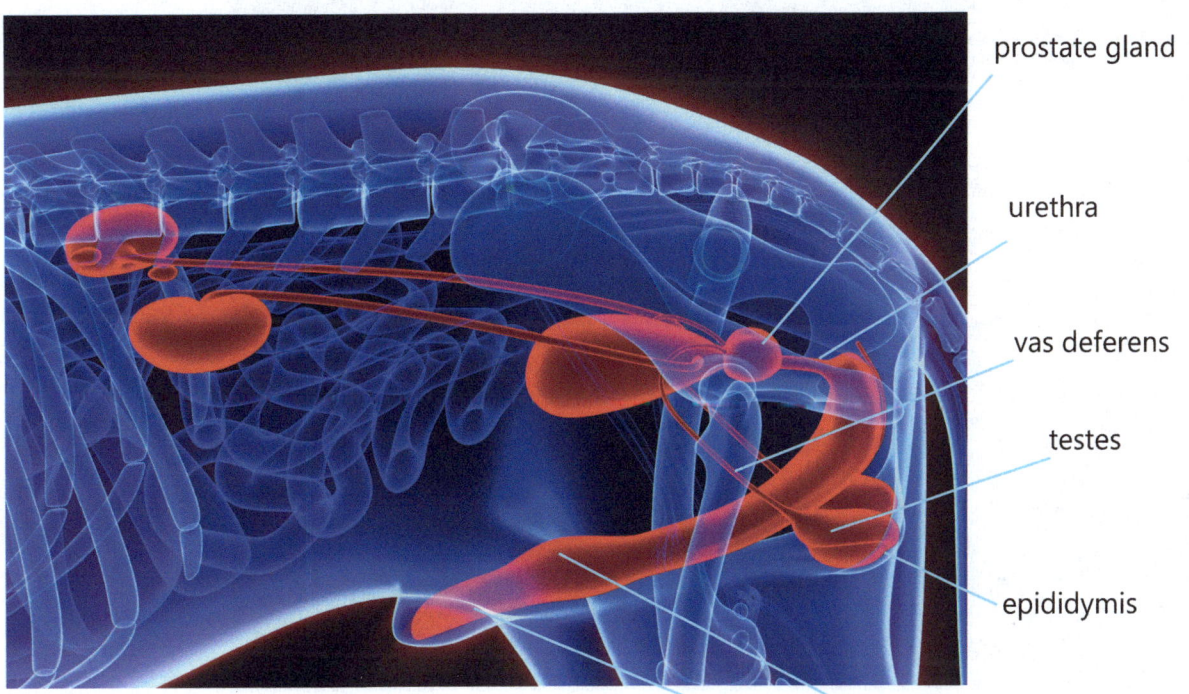

Figure 13 The male reproductive system

animals together, preventing them from separating until the bulb of erectile tissue retuns to normal.

- **Testes:** This pair of organs are responsible for the production of sperm. The testes produce **testosterone**, a hormone that plays a role in sperm production and secondary sexual characteristics. They also produce small amounts of the hormone, oestrogen. For most mammals the paired testes are located outside the body in the scrotum. ('Testes' is the plural of 'testis').

- **Epididymis:** A long coiled tube that connects the testes to the vas deferens. During transport through the epididymis the sperm mature, and mature sperm are stored here before release.

- **Vas deferens:** These tubes transport sperm from the epididymis to the prostate gland.

- **Prostate gland:** The vas deferens end at the prostate gland. Part of the urethra is enclosed by the prostate gland. It expels sperm into the urethra upon ejaculation and produces a fluid which protects them.

- **Urethra:** A muscular tube connected to the prostrate gland and bladder that runs the length of the penis and which transports urine and semen (sperm and seminal fluid) out of the body.

Female reproductive system

- **Vulva:** The external genitalia made up of folds of skin which protect the vagina from infection.

- **Vagina:** A tube made up of smooth muscle which is the site of penetration (and ejaculation for some species) during copulation.

- **Cervix:** Located between the vagina and the uterus this is a ring of smooth muscle (sphincter) that opens to allow sperm in and closes to stop infections entering the uterus. Most mammals have just one cervix but rats and some other species have two cervixes, each leading to a separate uterus.

- **Uterus:** This organ is where offspring develop from a fertilised egg through to birth. Fertilised eggs attach to the uterine lining to gain nourishment from the placenta. The uterus is made up of smooth muscle, to allow for the growth of offspring and to expel them at birth. Non-human mammals have a uterine body and uterine horns, which give more space for multiple offspring to develop.

- **Oviducts** (paired): Smooth muscle tubes which connect the ovaries to the uterus and where fertilisation of an ovum (egg) occurs. The tubes are lined with cilia which

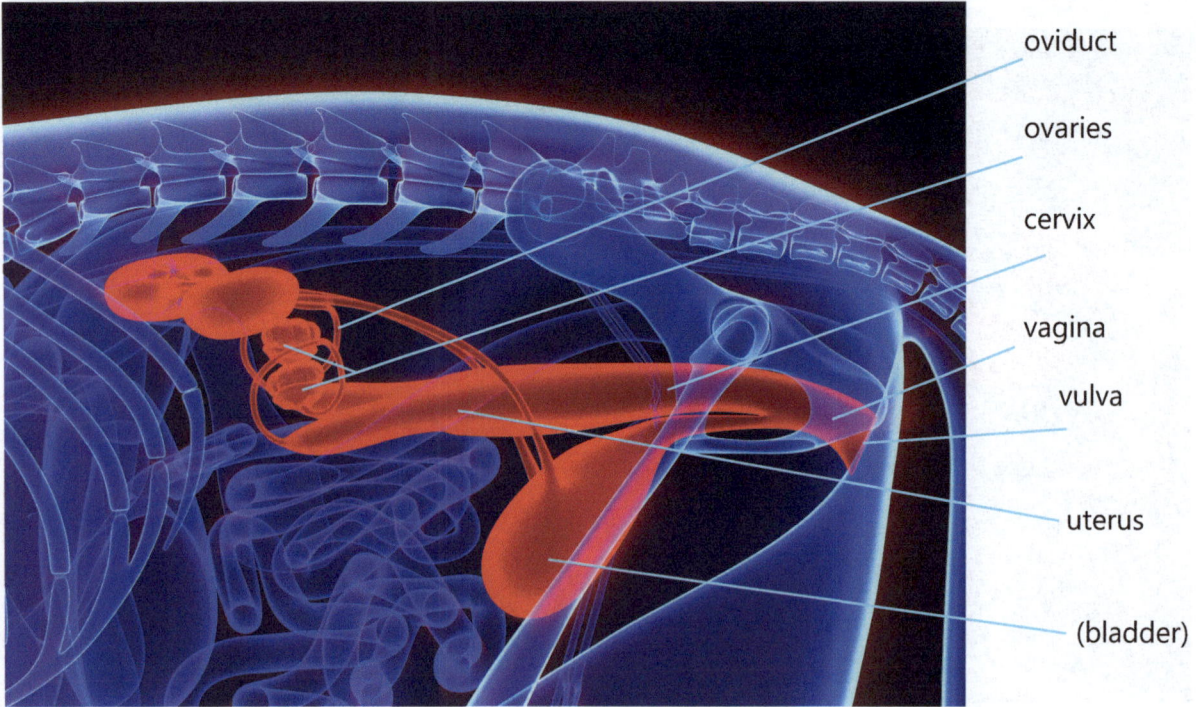

Figure 14 The female reproductive system

move the egg from the ovaries to the uterus. (Also known as fallopian tubes.)
- **Ovaries** (paired): Responsible for the production of an **ovum** (egg), which is released once it has matured. Ovaries are also responsible for the production of the hormones **oestrogen and progesterone**. Oestrogen plays a role in secondary sexual characteristics.

Adaptations of reproductive systems

Cats

In most mammals the erect penis points towards the animal's head. However a cat's erect penis points downwards and towards the animal's rear.

The cat's penis is also covered in hundreds of small barbs which stimulate the female during copulation to trigger ovulation (release eggs from the ovaries). In an induced ovulator, such as a cat, ovulation can only be triggered by mating. Most mammals are spontaneous ovulators, where ovulation occurs at a set point within the oestrus cycle (see next section) regardless of whether copulation has occurred.

Whales

Unlike other mammals, the reproductive organs of male whales are completely internal. This is to ensure that the whale is completely streamlined when swimming. The penis is curved in an S-shape inside the body, and only emerges through a genital slit during mating.

In other mammals the testes are held outside the body to keep them cool. This is because the high internal body temperature prevents the production of sperm. Whales face the same problem but they cool their internal testes using blood that has already flowed through the tail and fins. This blood is cool because it has been in contact with the coldest parts of the whale's body.

Unlike other female mammals, a whale's vagina has a series of complex folds and is not a straight tube. The precise reason for this is unclear but various suggestions include:

- to prevent seawater contaminating sperm after copulation
- to retain sperm and prevent it leaking out after copulation
- to counteract pressure changes caused by diving or surfacing, which could otherwise cause an embryo or foetus to be pushed from the uterus of a pregnant female.

Pigs

Female pigs (sows) have a heart-shaped uterus, with two long and winding uterine horns which connect to the oviducts. (The formal name for this is a bicornate uterus). The horns are much longer than the body of the uterus, and are the location for the foetuses during pregnancy. Their length means that a sow can support and give birth to large litters - an average of 12-14 piglets but sometimes as many as 20.

A sow's cervix has a series of interlocking folds. To allow entry, the end of the male pig's penis is shaped like a corkscrew. This means they are fully locked together during copulation, which typically lasts for 20 minutes. (Male pigs are known as boars.)

The shape of the sow's cervix means that plastic catheters, used for artificial insemination, are also corkscrew-shaped.

The oestrous cycle

The **oestrous cycle** is also known as the reproductive cycle. It describes the periods when a female mammal is fertile and sexually active, and when they are not. The timeframes for the cycle are different for each species but the key parts of the cycle are similar.

The cycle is mainly controlled by four **hormones**:

- **follicle stimulating hormone (FSH)** stimulate **follicles** to develop and mature. A fully mature follicle can release an ovum. FSH is released by the pituitary gland.

- **oestrogen** is secreted by the developing follicles (known as **Graafian follicles**) as they grow in the ovaries. It causes the lining of the uterus to thicken, in order to prepare it for copulation.

- **luteinising hormone (LH)** triggers the release of the ovum from the ovary, a process known as ovulation. It is secreted by the pituitary gland.

- **progesterone** is secreted by the remains of the follicle in the uterus, which is called the **corpus luteum**. Progesterone maintains the uterus lining. If fertilisation has occurred then progesterone levels remain high throughout pregnancy, which keeps the lining intact.

A part of the brain called the **hypothalamus** sends signals to the pituitary gland to begin the production of FSH, which in turn begins the whole cycle. The hypothalamus can be triggered by environmental conditions, such as the amount of daylight or temperature changes, which can regulate the natural cycle in some species.

Did you know?

When an animal is fertile they are referred to as being 'in oestrus' (or, less formally, 'on heat'). There are subtle spelling differences between the 'oestrous cycle' and being 'in oestrus'. To further confuse matters the American spelling drops the 'o' i.e. 'the estrous cycle', 'in estrus', (and also 'estrogen'). Don't be surprised to see these spellings when conducting research but be sure to use the British spelling, with the leading 'o', in your own work.

Link
Hormones, the pituitary gland and the hypothalamus are all discussed in more detail in section 2.1 The Endocrine System.

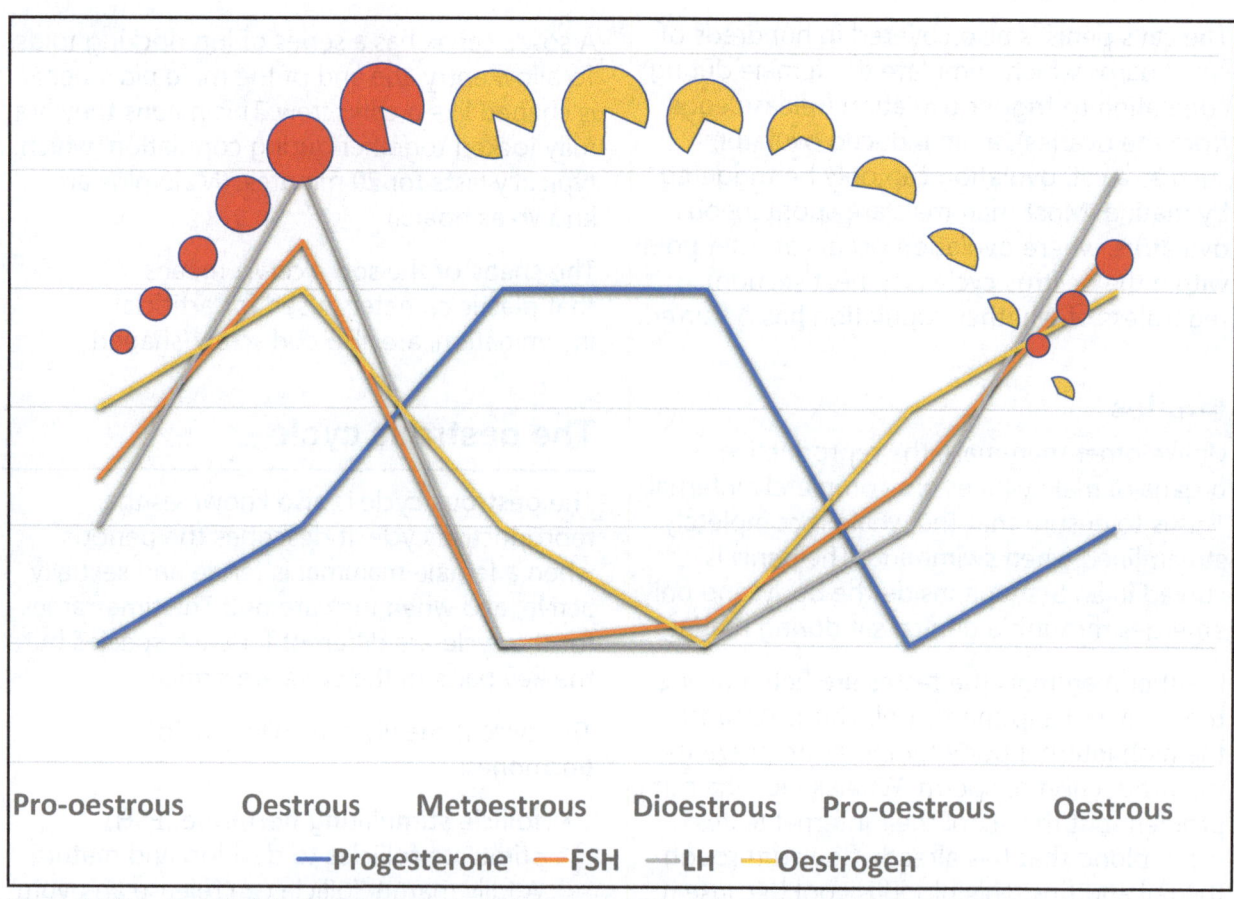

Figure 15 This simplified diagram represents the hormones present at each stage of the oestrous cycle. The red circles represent the growing follicles. The red segment represents the point of ovulation. The yellow segments represent the corpus luteum, which eventually degenerates, leading to the repeat of the cycle. The main points to note are that pro-oestrous and oestrus are known as the follicular stages, and are characterised by the presence of oestrogen. Metoestrous and dioestrous are known as the luteal stages and are characterised by the presence of progesterone. This is a simplified diagram. In reality the relative amounts of hormones have more complicated relationships, and will differ between species. For instance, LH is often secreted in pulses, whilst follicle development can occur in several waves rather than as one simple growth pattern as shown here.

Stage 1: Pro-oestrous

- The hypothalamus stimulates the pituitary gland to produce FSH and LH.
- FSH stimulates the growth and development of follicles in the ovary.
- The developing Graafian follicles in the ovary produce oestrogen.
- Rising oestrogen levels cause the uterus lining to form, ready to receive a fertilised egg.
- However, increased oestrogen inhibits further production of FSH.

Stage 2: Oestrous

- The Graafian follicles continue to mature, further increasing the production of oestrogen, which stimulates further production of LH
- LH triggers ovulation, where the ovum (egg) is released from the follicle and leaves the ovary.
- This is the fertile period, when the female is receptive to mating, evident by physical and behavioural changes.

Stage 3: Metoestrous

- The remains of the follicle that released the ovum is called the **corpus luteum.**
- Oestrogen levels decrease.
- The corpus luteum releases progesterone, which thickens the lining of the uterus to prepare it for a fertilised egg.
- Progesterone inhibits the production of FSH and LH.

Stage 4: Dioestrous

- If there is no pregnancy, the corpus luteum breaks down, decreasing progesterone levels
- Low levels of progesterone allow the production of FSH and LH, which starts the cycle again at pro-oestrous.

Anoestrous is when the cycle stops altogether and reproductive organs such as the ovaries have no activity. There are low levels of FSH, LH oestrogen and progesterone. It is often seasonal, occuring at certain times of the year.

Anoestrous occurs because the hypothalamus is influenced by the pineal gland. This gland produces a hormone called melatonin. Short, dark days cause more melatonin to be produced. In some species (e.g. horses) melatonin represses the hypothalamus from starting the oestrous cycle. In other species (e.g. sheep), melatonin stimulates the hypothalamus to begin the cycle.

Different species have different timeframes for each part of the oestrous cycle. Animals that are mono-oestrous only have one oestrous cycle per year. Poly-oestrous animals have more than one oestrous cycle per year. A species can have several cycles within a breeding season followed by an extended period of anoestrous.

Some typical oestrous cycle timeframes include:

- dog – two cycles per year
- cats – numerous cycles from spring to autumn
- horse – several cycles in spring and summer, but anoestrous in winter
- sheep – breeding season is August to February, so that lambs are born in spring when grass is abundant and has increased nutrients.

> **Did you know?**
>
> Most mammals have an oestrous cycle but only a few species have a menstrual cycle. The difference is that the lining of the uterus is absorbed by the body in the oestrus cycle and not discharged from the body as in the menstrual cycle. In addition, animals with an oestrous cycle are only sexually active in oestrus, whereas animals with a menstrual cycle are sexually active at any point in the cycle.

Human influence on breeding

The oestrous cycle is controlled by hormones. Professional breeders working with bitches and broodmares (female horse used for breeding) use hormone injections to control the animal's oestrous cycle. Examples of injections include:

- Progesterone, to increase the likelihood of a full-term pregnancy by helping the uterus lining to fully develop.
- Hormones that quicken the breakdown of the corpus luteum, which ends one cycle, so the next one can begin.
- LH to trigger ovulation to synchronise with insemination. 'Hormone sponges' are used for this purpose on some farms, to shorten the lambing period to a few days.
- Synthetic hormones which stimulate the hypothalamus, causing the production of FSH and LH to trigger ovulation.
- Hormones which prevent the oestrous cycle and stop animals from breeding. Medical procedures can also be used to permanently stop an animal from breeding.

Other ways in which humans can influence breeding include:

- Changing the environment to match the season that the species breeds in. For instance, changing the temperature, artificial light levels (to control day length) and nutrition to match a season such as spring.
- Using teaser animals. For instance, males who are sterile (unable to reproduce), but still producing sex hormones, can be placed into the same environment as females or allowed to walk nearby. Their presence can induce the oestrous cycle in some species, or help breeders to identify females who are in oestrus.
- Introducing male or female pheromones into the environment, to prepare animals for mating.

Sexual reproduction

There are several stages in the sexual reproductive cycle:

Copulation

- Also known as sexual intercourse or mating. In mammals the male's penis enters the female's vagina. Not all species have a penis - for example, birds do not.
- **Copulation** normally begins with courtship, where a female who is in oestrous gives a range of signals that she is ready to mate. Courtship differs between species.

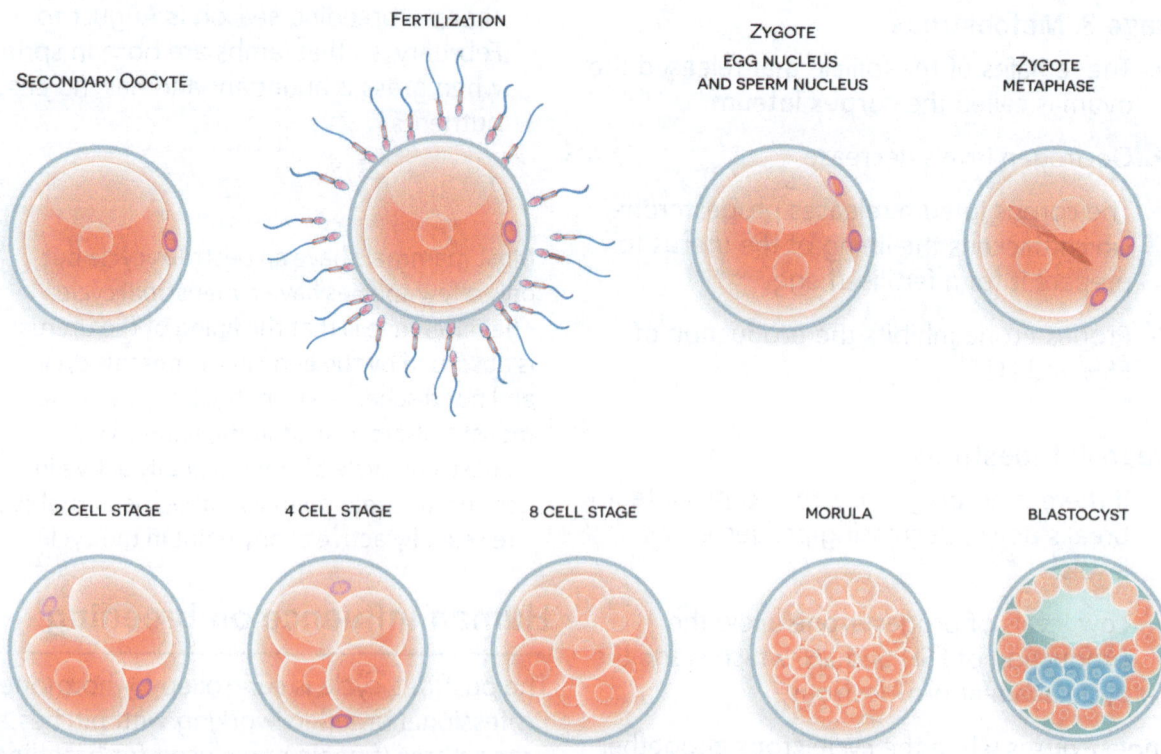

Figure 16 Fertilisation of an egg by a sperm

Insemination

This is where sperm enters the female's vagina. This can occur through copulation or through artificial insemination, where sperm is collected from the male and inserted separately into the female.

Fertilisation

This is when sperm fuses with an ovum.

- Semen contains millions of sperm, which travel through the cervix, and on through the uterus to the oviducts (fallopian tubes).

- At the same time, ovulation occurs - **ova** (eggs) are released from the ovaries and travel down the oviducts. The ova meet the sperm in the oviducts. Some species release just one ovum whereas other species release several ova to have multiple young.

- The oviducts contain the environmental conditions for **fertilisation** to occur. One sperm breaks through the zona pellucida, the outer membrane of an ovum, triggering a chemical change so no more sperm can enter that ovum.

- Once inside the ovum, the sperm loses its tail and the nucleus of the sperm fuses with the nucleus of the ovum.

- The nucleus of the ovum and the nucleus of the sperm each contain half the total number of chromosomes for the species. For instance, dogs have 78 chromosomes, which means their ova and sperm each have 39 chromosomes. The fusing of the two nuclei forms a new cell with a full set of chromosomes called a **zygote**.

- The zygote begins to grow by dividing (see Figure 16). It travels down the oviducts and into the uterus.

Implantation

- Whilst fertilisation has been taking place, the release of progesterone causes the uterus lining to thicken.

- As the zygote continues to grow it becomes a **blastocyst** (Figure 16) and implants into the uterus lining. The lining has an increased blood supply to provide the necessary nutrients for the further development.

Gestation

This phase covers the development of the young in the uterus from implanatation all the way to parturition (birth).

- Part of the blastocyst becomes the **embryo** itself, and part of it becomes the **placenta**.

- The cells of the embryo continue to multiply, with different cells creating different structures and providing different functions. The nervous system, brain and major organs are amongst the first to develop. The embryo stage is characterised by the creation of these major body systems. In the next stage the embryo becomes a **foetus**, where all organs, structures and systems grow and develop.

- At the end of the **gestation** period the foetus is fully developed and its various systems – such as the digestive or respiratory system – are all ready for life outside the uterus.

- As the foetus grows the mother's body changes - most notably the uterus expands and moves, causing the animal's abdomen to enlarge.

The gestation period varies for different species. Some average figures are:

- dog – nine weeks
- cat – nine weeks
- horse – 330 to 345 days
- rabbit – 30 days
- sheep – 147 days
- cow – 282 days

For comparison, a full-term human pregnancy lasts for 40 weeks (or 280 days/nine months).

Parturition

Parturition is the process of expelling the young from the uterus at the end of gestation. It is also known as 'giving birth'.

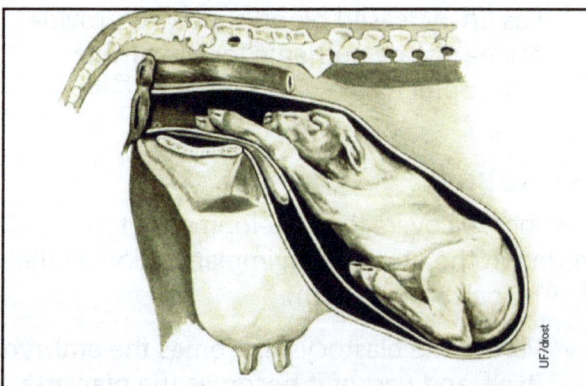

Figure 17 Normal presentation of a foetus during parturition. Reproduced with kind permission of Dr Maarten Drost.

- At the end of gestation, the offspring puts pressure on the cervix (Figure 17), releasing a hormone called **oxytocin** which causes contractions. Contractions are coordinated muscle movements in the uterus which begin to push the foetus (or foetuses) to the cervix and out through the vagina.

- Progesterone, produced throughout pregnancy, is a muscle relaxant which prevents contractions. At the end of gestation progesterone production falls.

- More pressure on the cervix causes more contractions, releasing more oxytocin which makes the contractions stronger and closer together. This pushes the offspring and placenta out of the female.

- For animals that give birth to a litter, the contractions continue until all of the offspring and placentas are out.

- The umbilical cord, which connects the offspring to the placenta, is broken by the mother after birth.

- In mammals the mother's body produces a hormone called **prolactin**, which is responsible for the production of milk to feed the offspring. Oxytocin is responsible for the muscle actions that push milk out through the nipples.

Adaptations

The reproductive characteristics covered so far relate specifically to mammals. Most mammals (including humans) are **viviparous**. Viviparous animals give birth to live young, and the embryos receive nutrients directly from the mother.

Non-mammals have some different characteristics – and some mammals have also adapted their reproductive systems. These are explained below.

Oviparous

In **oviparous** animals, embryos develop in eggs which hatch outside the body. Unlike viviparous animals the embryos do not receive nutrients directly from the mother, they receive them from the yolk. Examples include:

- birds
- most reptiles
- most amphibians
- most fish

Note that fertilisation of the egg can take place inside the body (e.g. birds, snakes) or outside the body (e.g. fish often lay eggs which a male fish then fertilises). The important point is that the young hatch outside the body.

Ovoviviparous

In **ovoviviparous** animals, embryos develop in eggs within the mother's body, which then hatch inside her before she gives birth. Unlike viviparous animals the embryos do not receive nutrients directly from the mother, they receive them from the yolk.

Figure 18 A short-beaked echidna

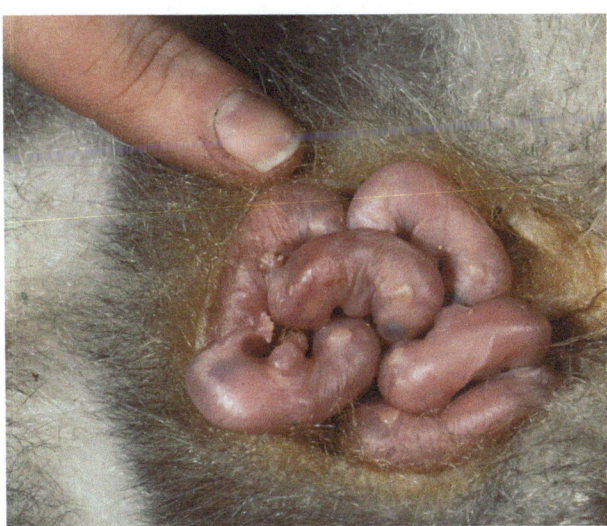

Figure 19 Opossums are marsupials. New born opossums (left) and when they are older (right)

Some examples of ovoviviparous animals are:

- boa constrictors
- stingrays
- Surinam toad

> **Link**
> Some further unusual characteristics of the duck-billed platypus are discussed in section 4.2

Egg-laying mammal

Almost all mammals give birth to live young. Viviparous mammals that follow the gestation patterns described previously are also known as **placental mammals**. However, a second type of mammal, called **monotremes**, lay eggs instead. There are ony five monotreme species: the duck-billed platypus, the short-beaked echidna (Figure 18) and three species of long-beaked echidna.

Monotremes are indigenous to Australia or New Guinea and split from other mammals around 200 million years ago. They have some reptile-like characteristics but share other traits with mammals:

- they are **endotherms** (see section 2.3) but have a lower body temperature than other mammmals
- they do not have teeth
- their limbs are on the side of their body, like reptiles, rather than underneath their body like other mammals
- they do not have nipples but they do produce milk for their young from openings in their skin (providing milk for the young is the definition of a mammal).

Marsupials

Marsupials are the third type of mammal (Figure 19). Their sexual reproduction cycle and anatomy is somewhat different to placental mammals. The key points are:

- A fertilised ovum becomes an embryo, embedded in the wall of the uterus, just like in placental mammals. However, a much less sophisticated placenta develops.

- The next stage of development, from embryo to foetus, is short and soon leads to birth. The newborn is far less developed than in placental mammals, but it makes its way to its mother's pouch after birth.

- Marsupials are famous for their pouches – a fold of skin in which they carry and nurse their young. Newly born marsupials continue their growth and development in the pouch rather than in the uterus, and drink their mother's milk instead of continuing to rely on nutrients from a placenta. They stay in the pouch for far longer than they did in the uterus.

- Marsupials have three vaginas, although they connect to one external opening, and two uteruses. The two outer vaginas are for sperm to travel up, and the inner vagina is used for birth.

Questions

1. What is the function of the ovaries in the female reproductive system? (2)

2. Label the following diagrams. (7)

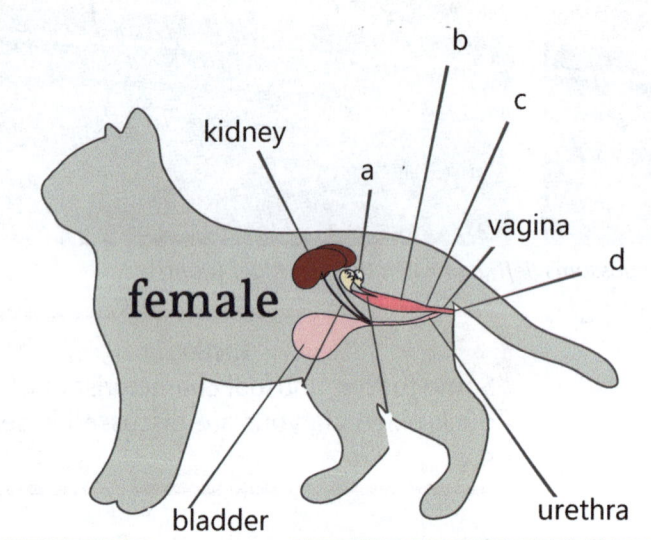

a _____

b _____

c _____

d _____

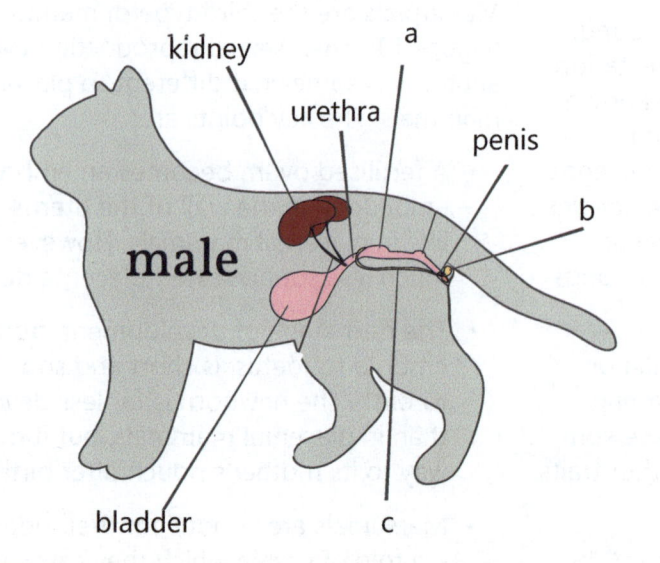

(Note: epididymis and bulbus glandus are not shown in this diagram)

a _____

b _____

c _____

3. a) State an adaptation to the reproductive system in a named species. b) What is the purpose of this adaptation? (4)

4. Outline the relative levels of oestrogen, progesterone, FSH and LSH in the four main stages of the oestrus cycle. (8)

5. Discuss the main events in the gestation stage of the sexual reproduction cycle in placental mammals. (4)

6. Explain how the sexual reproduction cycle in marsupials differs from placental mammals. (6)

1.4 Structure and function of the excretory system

In this topic you will learn about:

- The structure and function of the excretory system.
- Comparative adaptations of the excretory system in birds, desert mammals and aquatic animals.

The cells in animal bodies create waste products and toxins as part of their normal activities. These substances are removed from the body by the **excretory system**.

The excretory system is made up of the **kidneys**, **ureters**, **bladder** and **urethra**.

The main substances that are removed by the excretory system are water, salts and **urea**. These are combined together as urine.

Urea is a non-toxic substance that can be easily removed from the body. It is formed after processing **ammonia**. Ammonia is a toxic waste product, itself formed from the processing of amino acids, that must be removed from the body. The liver converts ammonia to urea and releases it into the bloodstream.

Oxygenated blood is delivered to both kidneys by the renal artery and filtered deoxygenated blood leaves the kidneys through the renal vein. It is dexoygenated because it supplies oxygen to the kidneys as it passes through them.

Kidneys

The kidneys have two main functions: **ultrafiltration** and selective **reabsorption**. The kidneys filter blood by separating it into different substances. Some of these substances are reabsorbed back into the blood and the rest is removed from the body in urine.

The kidneys are bean-shaped with a renal cortex, renal medulla and renal pelvis. Inside each kidney there are millions of nephrons. Nephrons are the site of ultrafiltration and selective reabsorption. A nephron is shown in Figure 20a. Each part is discussed below.

The **glomerulus** is a knot of blood capillaries surrounded by Bowman's capsule. Bowman's

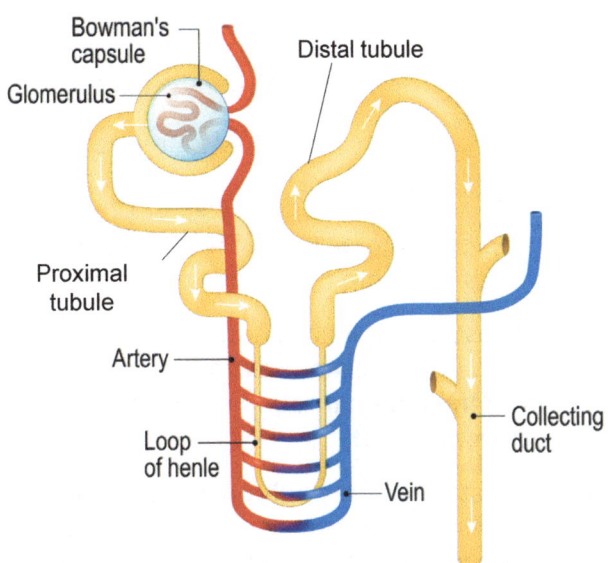

Figure 20a A nephron

capsule is made of epithelial tissue. The high pressure of the capillaries forces small molecules (such as water, urea, salt and glucose) into Bowman's capsule but large molecules stay in the blood. This process is called ultrafiltration.

The filtrate (all the small molecules that have been filtered out of the blood) moves into the **proximal tube**. Here, all of the glucose and around 65% of the water and salt are reabsorbed back into blood capillaries that surround the nephron. This is called selective reabsorption because not everything is reabsorbed back into the blood capillaries.

The filtrate moves into the **Loop of Henle**:

- The descending limb is permeable to water, which is removed from the filtrate through osmosis. The salt concentration of the filtrate increases down the descending limb, as more water is removed. The longer the Loop of Henle the more water can be reabsorbed back into the blood. Within the

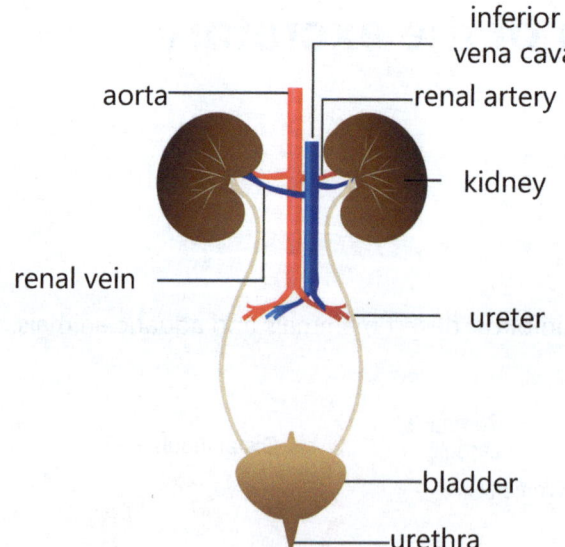

Figure 20 The excretory system

Loop of Henle, around 25% of the water moves back into the blood capillaries.

- The ascending limb is non-permeable to water. In this section salts are removed from the filtrate through diffusion and active transport.

The next stage of the nephron is the **distal tubule**, which selectively reabsorbs around 5% of the water and salt back into the blood capillaries.

The final stage is the **collecting duct**, which takes urine towards the ureters. If anti-diuretic hormone (ADH - see pages 48-50) is present then pores in the tube open, allowing water to return to the blood when the animal is dehydrated.

The hypothalamus detects how much water and salt is in the blood. If the animal is dehydrated, ADH is released and more water is returned to the blood within the collecting duct of the nephron. This conserves water in the body. If there is an excess of water in the bloodstream, no ADH is released, and water passes through the collecting duct and is removed in urine. This process of controlling the balance of water and salts in the body is called osmoregulation. This is an example of **homeostasis**.

The balance of salts in the blood, maintained by the kidneys, is important for the repair and maintenance of bones.

> **Link**
> For more on homeostasis see section 2.1. See section 1.5 for more on the bones and skeletal system.

Ureters

The kidneys pass urine through the **ureters**, a pair of hollow tubes that attach to each kidney. The ureters transport urine from the kidneys to the bladder.

Bladder

The bladder collects, temporarily stores and releases urine. The bladder is a balloon-like organ that is able to stretch to temporarily store urine. The lining of the bladder has folds which allow the bladder to expand. It has stretch receptors in the lining to detect when the bladder is full, which send messages to the brain to release the urine within it. There are muscles in the neck of the bladder that hold it closed until the urine is ready to be released.

Adults normally have voluntary control of the bladder. When it is full the brain sends messages to muscles in the neck of the bladder to allow it to empty. Adults can control this, to allow them to empty their bladder at a suitable time.

However, not all animals have control over this part of the brain. This means they have involuntary control of the bladder. Problems with the nervous system or brain can affect an animal's control of the bladder, and older or injured animals may not have full bladder control. Young domesticated companion animals need to learn how to control their bladder.

Although the bladder can expand, it is limited in size, so once completely full it will be emptied involuntarily, even if an animal normally has voluntary control.

Some species do not have voluntary control of their bladder - notably rats and mice. This is why they can be a source of contamination and spread disease.

Urethra

A single tube that connects the bladder to the external sexual organs. It transports urine from the bladder to the outside of the body. In males it also transports seminal fluid out of the body.

Adaptations

Birds

- Birds' kidneys process ammonia into **uric acid** rather than urea.
- Uric acid is even less toxic than urea, which means it can build up within eggs without damaging the embryo
- Converting ammonia to uric acid conserves more water, but requires more energy, compared with converting ammonia to urea.
- Birds do not have a bladder - instead they only have one chamber and orifice for both liquid and solid waste, as well as reproductive functions, called the **cloaca**.
- Faeces and uric acid are excreted from the cloaca as one semisolid paste.

Desert mammals

Desert mammals, such as camels and kangaroo rats, can survive on small amounts of water that would be fatal for other mammals. To do this they need to conserve as much water as possible. Their excretory system has adapted so that it can:

- reduce the rate of ultrafiltration
- reabsorb more water due to a longer Loop of Henle

The Loop of Henle is 4-6 times longer than in cattle, which returns more water to the bloodstream. Very concentrated urine is produced and very little water is lost but other waste products are still removed. This adaptation also allows desert mammals to drink salt water that would be toxic to other mammals.

In addition, desert mammals like the kangaroo rat produce their own water by oxidising

Figure 21 A kangaroo rat

energy in their food. A large percentage of their water requirements are met in this way.

Aquatic animals

Aquatic animals also have to remove ammonia from their body. However, unlike mammals:

- most aquatic animals, including most fish, remove ammonia directly
- they can do this because ammonia dissolves in water, and aquatic animals are surrounded by water
- this allows ammonia to be quickly removed from the body
- fish use their gills to remove ammonia
- fish also produce urea which is extracted by the kidneys in urine - but most ammonia is removed directly via the gills.

The concentration of salts in the body is crucial to life. Too high or too low a concentration of salts is fatal. Aquatic animals are surrounded by water with a higher or lower concentration of salts than in their body. Many aquatic animals such as fish have skin that is permeable to water. This means water will pass through their skin in a process called osmosis which is shown in Figure 22. In osmosis, a permeable barrier separates water with a low concentration of salts and water with a high concentration of salts. The permeable barrier allows water to flow through it but prevents the salts from passing through it. This difference in concentration causes water to flow through the barrier, until the concentrations are the same on each side. In the first diagram in Figure 22 ('Before'), the left hand side of the barrier has a lower concentration of salts. Water flows from left to right to make the concentrations equal on both sides, as show in the second diagram ('After').

Fish can ether live in fresh water (e.g. rivers, lakes) or in salt water (e.g. oceans, seas and estuaries). Freshwater fish are surrounded by water with a lower concentration of salts than in their body. In this case, the right hand side of Figure 22 is like the fish's body, the barrier is the fish's skin, and the left hand side is the lake or river.

- There is a flow of water into the fish from the lake or river, to try and equalise the salt concentration. The fish is constantly gaining water.
- This has the effect of lowering the concentration of salts in the body to dangerously low levels.
- So, to prevent this, the kidneys of most freshwater fish are adapted to remove lots of water from the body, with large amounts of very dilute urine.
- Freshwater fish rarely need to drink the surrounding water because they are constantly removing water and trying to preserve the concentration of salts.

By contrast, saltwater fish are surrounded by water with a very high concentration of salts. In this case, the left-hand side of Figure 22 is like the fish's body, the barrier is the fish's skin, and the right-hand side is the sea or ocean.

- There is a flow of water from the fish into the sea or ocean, to try and equalise the salt concentration. The fish is constantly losing water.
- This has the effect of increasing the concentration of salts in the body to dangerously high levels.
- So, to prevent this, the kidneys of most saltwater fish are adapted to remove lots of salts from the body, with small amounts of concentrated urine.
- At the same time, saltwater fish constantly drink the surrounding salt water. Even

OSMOSIS

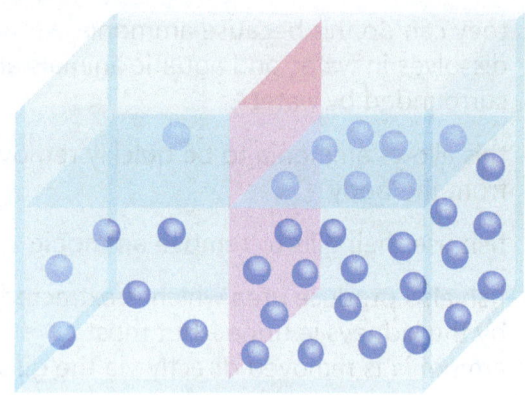

BEFORE

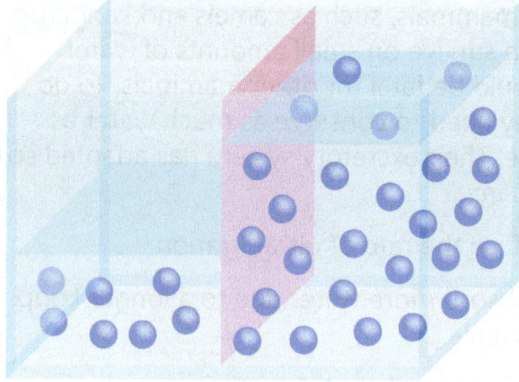

AFTER

Figure 22 The process of osmosis. In this diagram salt molecules are shown by the blue spheres. BEFORE: a permeable barrier separates water with a low concentration of salts (left) and water with a high concentration of salts (right). The permeable barrier allows water to flow through it but prevents the salts from passing through. This difference in concentration causes water to spontaneously flow from the left to the right, through the barrier, until the concentrations are the same on each side. AFTER: the water on the left has flowed to the right, and now the concentration of salt molecules is the same on both sides of the membrane, which stops any further flow of water.

though this means they are taking in more salts it also means they are taking in more water, and then their excretory system can then extract and remove the salts. This preserves the concentration of salts.

Some aquatic animals, such as sharks, crabs and octopus, have almost exactly the same salt balance in their body as in the surrounding water. This means their excretory systems do not have to work so hard maintaining salt concentration levels, because there is little or no water movement across the skin due to osmosis

Sharks maintain their salt balance by producing and storing some urea, which increases the concentration of salts in their body to match the surrounding water. This is unlike other saltwater fish, which produce very little urea.

Questions

1. Label the following diagram. (4)

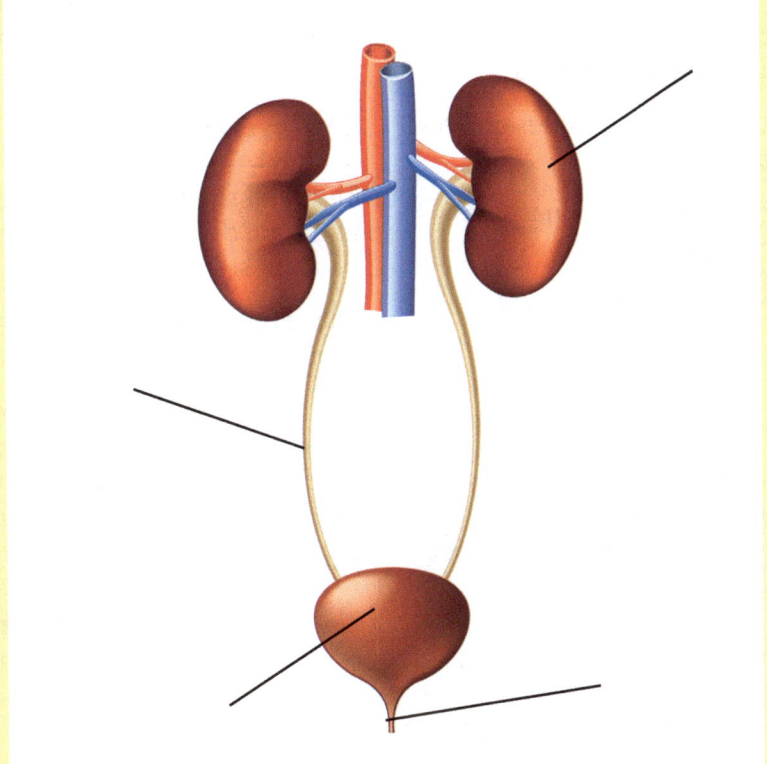

2. Describe the functions of each of the organs in the diagram. (5)

3. a) What toxin does the excretory system remove from the body? b) With reference to a named species, discuss three different ways in which animals get rid of this toxin. (4)

4. Give an example of an animal with involuntary control of their bladder. (1)

5. Explain how the structure of the bladder aids its function. (3)

33

1.5 Structure and function of the musculoskeletal systems

In this topic you will learn about:

- The structure and function of the mammalian musculoskeletal system.

- The adaptations that resulted in the mammalian musculoskeletal system.

- Adaptations to the mammalian musculoskeletal system in aquatic mammals, flying mammals, hopping mammals and running mammals.

The musculoskeletal system refers to the structure and function of the muscles and skeleton.

- The skeleton is made up of bones.
- **Ligaments** attach bones to bones and also stabilise joints.
- **Tendons** attach muscles to bones and allow flexibility within the joint.
- Muscles contract and relax to allow movement.
- Synovial fluid acts like a lubricant to help the joints move and reduce friction.

All mammals share the same basic features, which evolved from more primitive ancestors and allowed mammals to become successful.

The skeleton has five functions:

- **Protection**. The hard skeleton protects soft organs, such as the skull protecting the brain and the rib cage protecting the heart and lungs.
- **Structure**. It provides a rigid structure with attachments for ligaments and tendons.
- **Shape**. The skeleton gives shape to the animal.
- **Movement**. Working with the muscles, the skeleton allows the body to move
- **Storage**. Bones store calcium which can be released into the bloodstream when needed.
- **Production**. Bone marrow, a soft tissue found in the centre of some bones, produces erythrocytes, leukocytes and platelets (see section 1.1).

There are two main parts to the skeleton: the axial skeleton and the appendicular skeleton.

The axial skeleton

The **axial skeleton** protects important organs and is made up of the skull, vertebral column, ribs and sternum:

The **skull** protects the brain. Mammals have proportionally larger skulls than non-mammals.

The **vertebral column** (also known as the spine or backbone) is made up of **vertebrae** which enclose and protect the spinal cord. The number of vertebrae varies in different species.

The first two vertebrae have special names:

- The first vertebrae in the neck is called the **atlas**. It allows the head to move up and down.
- The second vertebrae in the neck is called the **axis**. It allows the head to move side to side.

The vertebral column is split into different sections, as follows:

- Cervical: The vertebrae in the neck. This section supports the head and neck and includes the atlas and axis.
- Thoracic: The vertebrae in the upper torso, with attachment points for the ribs. This section allows for some small movements to aid breathing.
- Lumbar: The lower back which allows for movement forwards and backwards as well

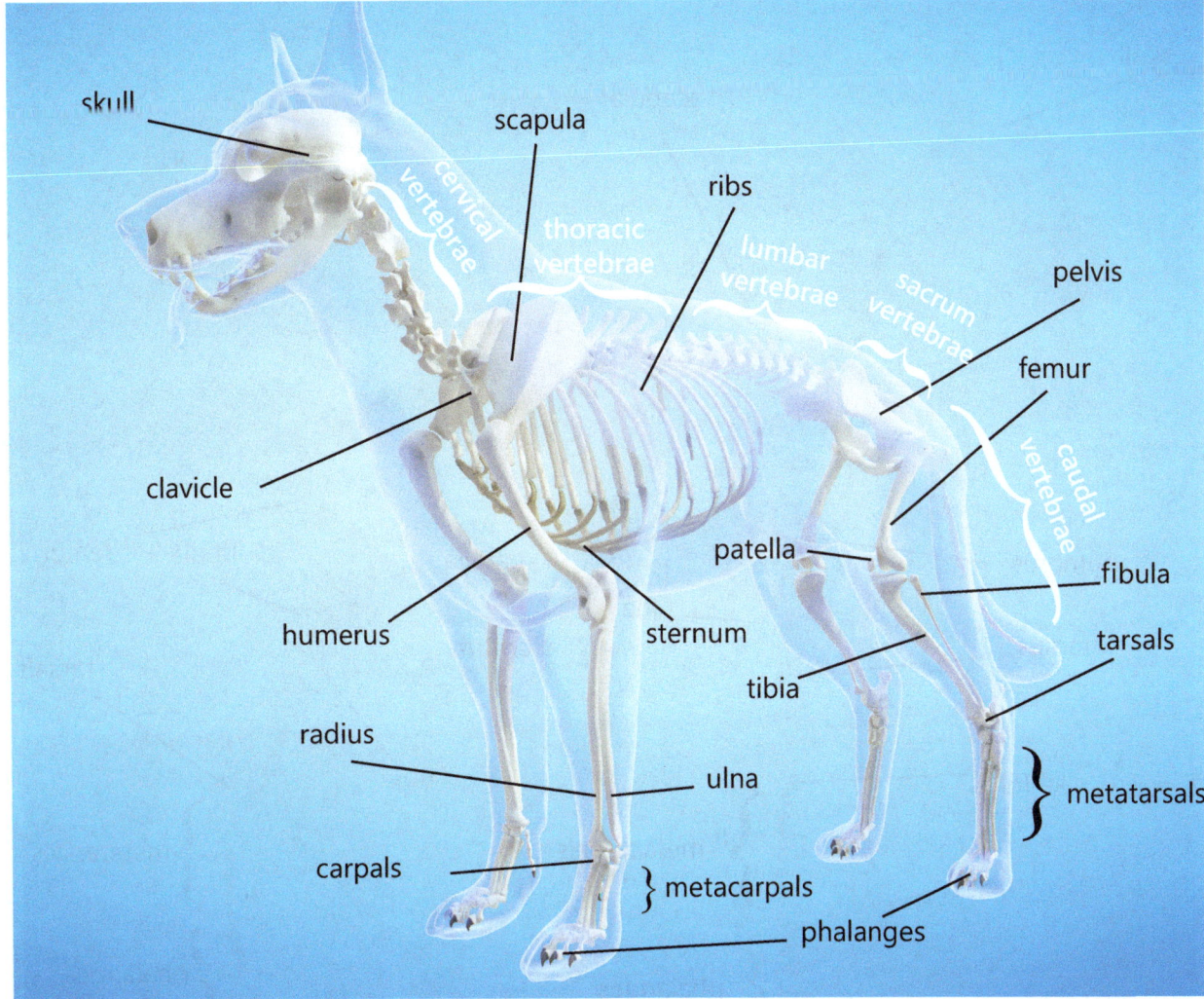

Figure 23 A dog's skeleton

as side-to-side. It connects to the pelvis.
- Sacrum: Vertebrae covered by the pelvis.
- Caudal/Coccygeal: The vertebrae in the tail which give the tail structure.

The **ribs** and the **sternum** (breastbone) together form the rib cage. The rib cage encloses and protects the lungs and heart, and provides the space into which the lungs expand when breathing.

The appendicular skeleton

The rest of the skeleton is called the **appendicular skeleton**. It is attached to the axial skeleton. It is made up of four limbs and the bones which support them.

The forelimbs are the front limbs. Proceeding down the forelimb:

- The upper bone is called the **humerus**.
- There are two bones in the lower limb, called the **radius** and **ulna**.
- The bones in the wrist are called the **carpals**.
- The **metacarpals** connect the carpals to the phalanges.
- The **phalanges** are the bones in the toes of the forelimbs and hindlimbs.

In some species the forelimbs are further supported by the **scapula** (shoulder bone) and the **clavicle** (collar bone). However, species adapted for running on all four legs tend to have a reduced clavicle that does not provide support, or no clavicle at all.

The hindlimbs are the rear limbs. Proceeding down the hindlimb:

35

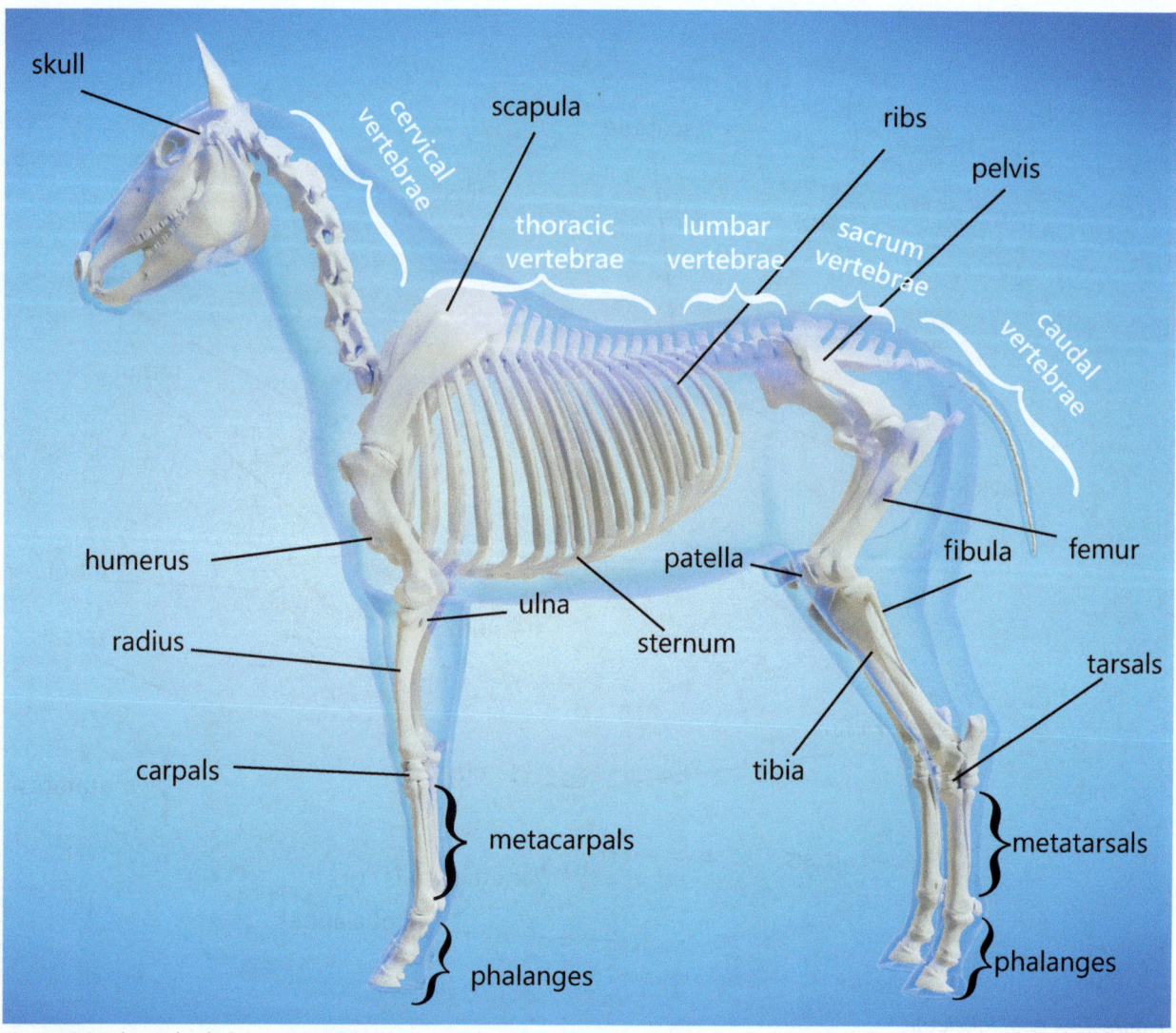

Figure 24 A horse's skeleton. Note that a horse does not have a clavicle, and the radius and ulna are fused together for most of their length, to provide stability when running. Also notice that a horse walks on its phalanges – see Figure 28 for the equivalent bones in humans.

- The upper bone is called the **femur**.
- The patella is the kneecap, on the joint between the femur and tibia / fibula
- There are two bones in the lower limb, the **tibia** and **fibula**.
- The bones in the ankle are called the **tarsals**.
- The **metatarsals** connect the tarsals to the phalanges.
- The phalanges are the toes.

The hindlimbs are supported by the **pelvis**. The pelvis is made up of the **sacrum**, **pubis**, **ilium** and **ischium**.

Bones can also be classified by their shape:

- **Long bones** are cylindrical in shape. They do not actually have to be long, they simply have to be longer than they are wide. Examples of long bones are the humerus, radius and ulna, and the femur, tibia and fibula.
- **Short bones** are roughly cube-shaped, being equal in all dimensions. Examples include the carpals and tarsals.
- **Flat bones** are thin though not necessarily flat – they are often curved. Examples include the bones that make up the skull, the scapula, the sternum and the ribs.
- **Sesamoid bones** have a similar shape to sesame seeds. They are sometimes called floating bones as they aren't attached to other bones but are found within tendons.

They help with the movement of tendons over joints. The patella is a sesamoid bone. Sesamoid bones are also found in the tendons of the paws.

- **Irregular bones** are those that don't fit into the classifications above. Examples include vertebrae, the lower jaw (mandible), a number of other bones in the skull that are not flat bones, and the bones of the hip – the pubis, ilium and ischium.

Muscular system

There are three types of muscle in an animal's body:

- smooth muscle
- cardiac muscle
- skeletal muscle.

Smooth muscle is mostly found in internal organs such as the stomach and intestines. It is spindle-shaped and not under voluntary control – instead it is controlled by the autonomic nervous system (see section 3.2). Here are some examples:

- Smooth muscle pushes food through the digestive system by contracting the walls of the stomach and intestines.
- Smooth muscle contracts the walls of the bladder to expel urine from the body.
- The smooth muscle of the uterus allows the uterus to stretch, so the foetus can develop and grow, and also pushes offspring out of the body by contracting during parturition.

> **Link**
> For more on the digestive system see Unit 304 section 1.3. For more on the function of the bladder see section 1.4. For more on parturition see section 1.3. For more on the autonomic nervous system see section 3.2.

Cardiac muscle is only found in the heart. It is not under voluntary control. Unlike other types of muscle, it never gets tired. This is because its cells include a large number of mitochondria, far more than in other muscle types. Mitochondria are responsible for producing the chemical energy that muscles need to keep working. Cardiac muscle is striated (striped) and muscle cells are "Y"-shaped so they can form branches with adjoining cells to synchronise the heartbeat.

Skeletal muscles attach to the skeleton with tendons and are under voluntary control (except for the diaphragm). Each muscle contracts to produce movement in one direction. Movement in the opposite direction requires another muscle to contract. Therefore, muscles work in antagonistic pairs to create a full range of body movements. Skeletal muscles are also striated, with lots of mitochondria (but fewer than cardiac muscle) and each muscle cell has multiple nuclei.

Joints are the points where bones meet. Synovial joints allow movement in different directions between the bones. (There are other joints that allow very limited movement, or no movement at all). Different synovial joints allow movement in different directions. They are classified according to their shape in the Nomina Anatomica Veterinaria (www.wava-amav.org/index.html) as follows:

- Spheroid joint (also known as a ball and socket), which allows up and down, and side-to-side movement, plus rotation e.g. the shoulder or hip.
- Hinge joint, which allows up and down movement e.g. the elbow or knee.
- Pivot joint, which allows rotation e.g. the between the atlas and axis vertebrae in the neck, which allows the head to turn.
- Plane joint (also known as a gliding joint), which allows side-to-side movement e.g. between the carpals.
- Condyloid joint, which allows up and down and limited side-to-side movement e.g. the wrist joint between the radius and carpals.
- Saddle joint, which allows a wide range of up and down and side-to-side movement e.g. in primates between the thumb metacarpal and the carpals.

The mammal muscular system is similar to that of reptiles. One major difference is the diaphragm, which is unique to mammals and allows them to breathe.

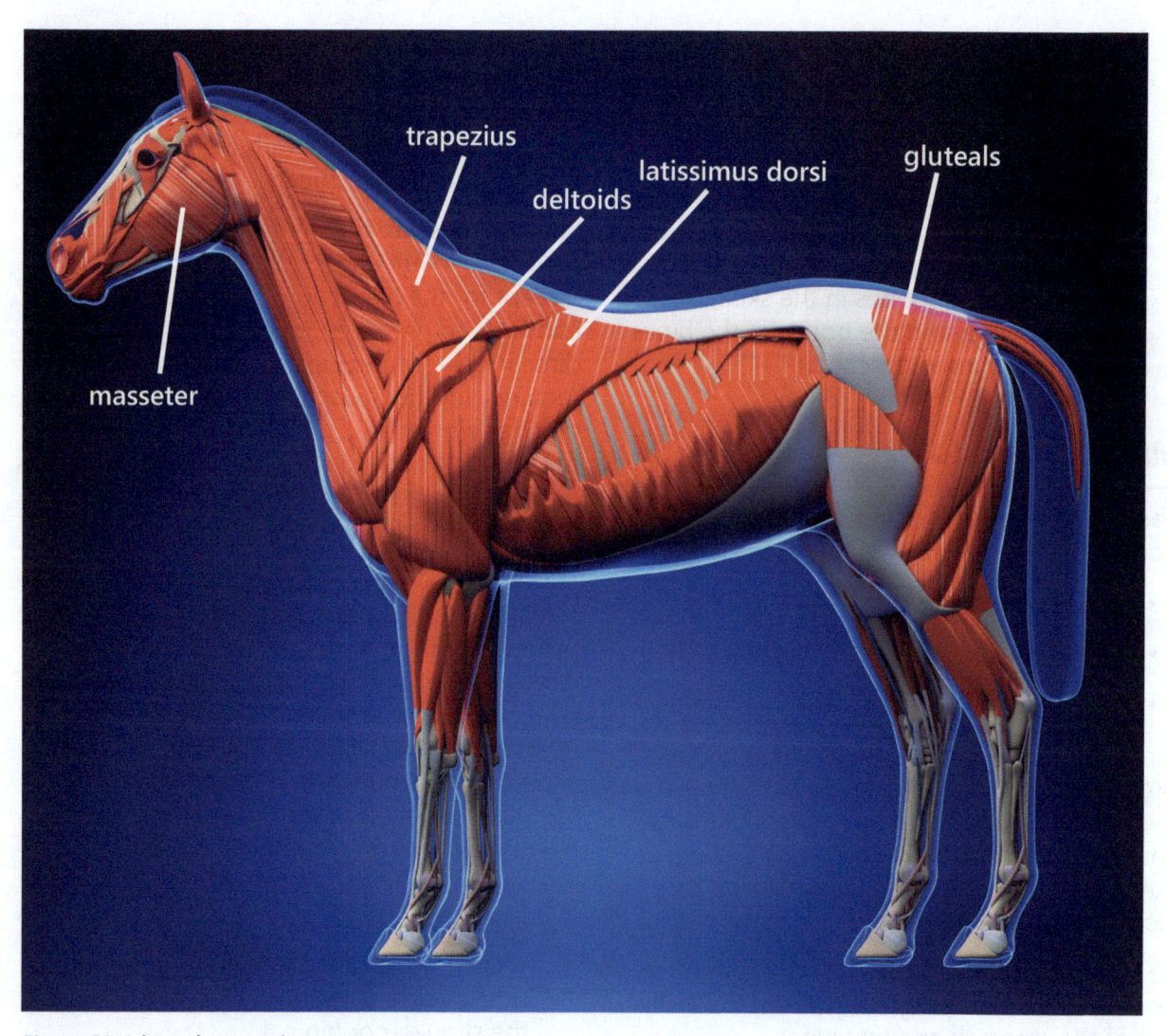

Figure 25 A horse's muscular system

Whilst there are hundreds of different skeletal muscles, which must all work together to provide a normal range of movement, some of the major muscles include:

- latissimus dorsi – large muscle across the back that helps shoulder movement
- deltoids – controls movement in the shoulder
- gluteals – a series of muscles that control movement of the hip and keeps it stable
- **diaphragm** – the muscle that facilitates the mammalian respiratory system
- masseter – this muscle in the jaw of mammals is important for chewing and biting

A (very) brief history of mammalian evolution

- Around 300 million years ago, there emerged two broad classes of four-legged reptiles called synapsids and sauropsids.
- The defining feature of the synapsids, compared to sauropsids, was a single opening in the skull behind each eye. The top of the opening is an arch (which is commonly known as the cheekbone in humans).
- The synapsids dominated for around 50 million years but there was a mass extinction event around 250 million years ago; this event wiped out many synapsids, and led to a new era dominated by sauropsids. The most famous example of saurospids, that came along another 50 million years later, were the dinosaurs.

Figure 26 An artist's impression of a lycaenops, a type of carnivorous therapsid, which lived around 250 million years ago. Therapsids are the ancestors of modern mammals.

- One group of synapsids that did survive the extinction event, however, were called the therapsids (sometimes known as 'mammal-like reptiles', Figure 26). They were small and much less dominant than those synapsids that had been wiped out; however, all mammals are descended from therapsids.

- Another extinction event 65 million years ago wiped out the dinosaurs and many other species too. By this point true mammals had evolved, and those that survived could grow, flourish and multiply in the absence of the previously dominant dinosaurs.

- Mammals are the only modern remaining examples of synapsids, with the opening in the skull behind the eye.

- Dinosaurs, and all modern birds and reptiles, are examples of sauropsids.

> **Link**
> There is more coverage of the principles of animal evolution in Unit 307 section 1.1

Mammalian musculoskeletal adaptations

- The synapsids' defining feature, the opening in the skull behind the eye, is thought to have provided space for larger jaw muscles to develop.

- In contrast to reptiles, the lower jaw in mammals is made of only one bone, with two muscles which close the jaw. This allows the side-to-side movement necessary for chewing. Chewing is almost unique to mammals, and was a desirable evolutionary development because it allowed for quicker, more efficient digestion of food. This was important to support the development of mammalian characteristics, such as being warm-blooded with a relatively high energy requirement.

- As mammals evolved, the lower jaw also grew larger, which allowed it to support more teeth with specialised functions, to further aid eating and digestion.

- Mammals developed a hard palate which separates the nasal passages from the mouth cavity. Part of the hard palate is formed from bones that evolved in the skull. A hard palate allows mammals to eat and breathe at the same time, which is partcularly important for the young suckling milk. Most reptiles do not have a hard palate – instead their nostrils open directly into their mouth.

- A mammal's middle ear has three tiny bones. This is unique to mammals. Two of these evolved from bones that make up a reptile's jaw. The change in structure, function and position of these two bones is an example of evolution. Three middle ear bones (the malleus, incus and stapes) allowed mammals to develop much more sensitive hearing than reptiles. It is likely that the earliest mammals were small, nocturnal creatures, for whom good hearing would have been vital for detecting prey and avoiding predators.

> **Link**
> For more on the bones of the mammalian ear and the palate see section 4.1.

- Most mammals' limbs are underneath the body, in contrast to reptiles who have limbs at the side of the body (Figure 27). This evolutionary change allows for more efficient running and walking, and required changes to the pelvis, scapula and clavicle.

Figure 27 A komodo dragon. Notice that a reptile's legs are mounted on the side of the body, in contrast to most mammals.

- This change in limb position led to a more upright posture, which is potentially more unstable. Mammals' tails became shorter, with fewer vertebrae, to aid with balance.

- The development of a heel bone (calcaneal) and a larger calf muscle allows mammals to push off from the ground – which partly accounts for various species' abilities to run fast and jump high.

- The diaphragm is a muscle that is unique to mammals. It is a smooth muscle that is not under voluntary control. As discussed in section 1.2, the diaphragm allows the lungs to expand and contract, creating a much more efficient respiratory system than in reptiles. This allows mammals to supply more energy to their cells, which allows them to be more active. (It should be noted, however, that birds evolved an even more efficient respiratory system, that does not require a diaphragm.)

- Reptiles have a rib cage that extends all the way to their tail. Lizards use the same muscles for breathing and running so they cannot do both at the same time. Mammals lost the lower ribs during evolution, providing space for the development of the diaphragm, space for developing embryos, and to allow for more flexible movement of the hind legs.

Injury and disease of the musculoskeletal system

Osteoarthritis

This condition is due to inflamed joints, leading to pain, stiffness and lost mobility. It is caused by damaged or worn cartilage between joints, and tends to affect animals as they get older.

Osteomalacia

An imbalance of minerals in the diet (such as calcium and phosphorus) which leads to soft bones, causing pain and problems with posture and movement.

Tendon and ligament injuries

Large forces are transmitted through tendons and ligaments. If they are overstretched, or forced to move in the wrong direction (e.g. after awkwardly planting a foot), they can tear. Such tears are painful, take a long time to heal, and lead to reduced movement or ability to walk.

Broken and fractured bones

Some bones are more prone to breaking and fracture than others, and the details will depend on species - but the bones of both sets of legs are often most prone. Whilst this is rarely life-threatening in companion animals or livestock, in the wild a broken bone can lead to death due to starvation (unable to catch prey or gather food effectively) or predation (i.e. unable to escape a predator).

Adaptations within mammal species

Although all mammals share certain musculoskeletal features, some species have developed further adaptations as a result of their environment and lifestyle.

Aquatic mammals

Aquatic mammals such as cetaceans (whales, dolphins and porpoises), evolved from land-dwelling mammals. Many aspects of the

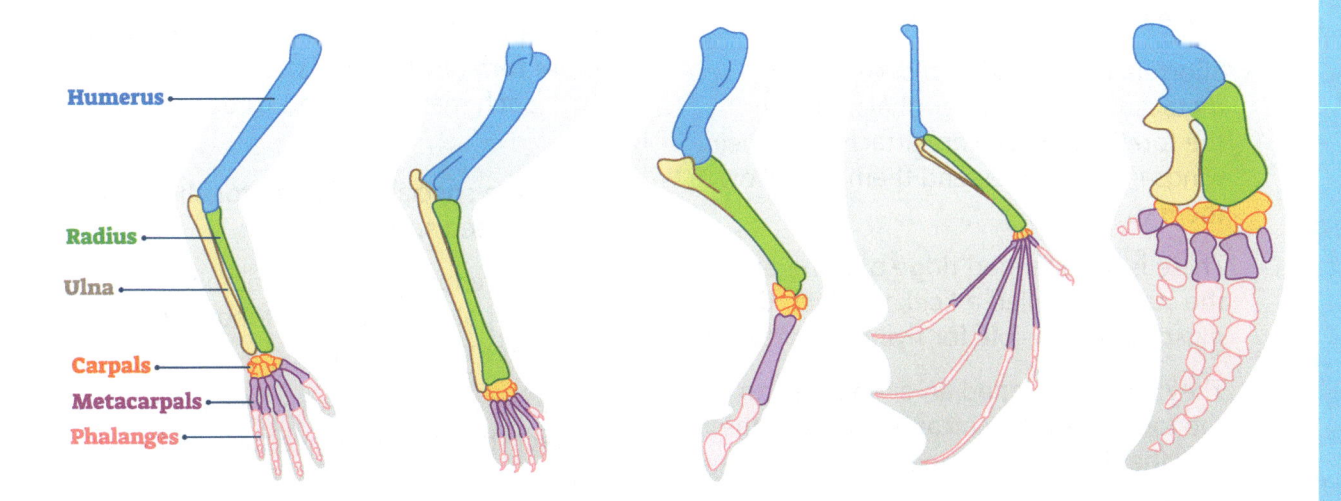

Figure 28 From left to right: human, cat, horse, bat, dolphin. It is interesting to note that each of these very different mammals has the same bone structure.

skeleton are the same – for instance cetaceans retain a humerus, radius, ulna and phalanges within each flipper. However, the demands of living in water have resulted in a number of changes to cetaceans:

- The skeletal bones are lighter, because being in water means they do not need to support the full mass of the animal.
- Much shorter humerus but longer phalanges, to allow the flippers to act more like paddles which are used for steering.
- There are only four 'fingers' but the phalanges are much longer than the metacarpals, in contrast to land mammals.
- They do not have hindlimbs – although the initial development of very small hindlimbs can be seen in embryos. They do not need them because they use their tale to power through the water and are more streamlined without them.
- Whales have a vestigial pelvis. This is a floating pelvis as it is not attached to the spine, and without any hindlimbs it is no longer needed.
- Compressed or fused neck vertebrae, to withstand forces on the neck from diving and swimming, which has resulted in a very short neck.
- Greater flexibility of vertebrae, to allow for greater movement and more powerful thrusts, and a greater number of tail vertebrae and muscles, as the tail provides most of the power for swimming.
- Whales' ribs can adjust position, to withstand high pressures when diving by collapsing their lungs. But the drawback is that the ribs are unattached to a sternum so if they ever leave the water they are crushed by their own weight. Without the water to support their body weight they suffocate.
- Much longer ('telescoped') skull – more streamlined for efficient swimming and better adapted for feeding in the water.
- Nostrils are located at the top of the skull, seen as blowholes.

Flying mammals

Bats are the only mammals that can truly fly rather than glide. To achieve this there are a number of musculoskeletal adaptations:

- The skeleton is made up of thin bones, to keep bats as light as possible, but their bones are actually more dense than similarly sized non-flying mammals – this helps them to withstand the forces associated with flying. But the drawback is that the lighter bones are more prone to breaking.
- The first phalange is small and clawed which is used by the bat to climb or to walk.

- Some bones in the skull are fused, to help make the skeleton lighter.
- Shortened ulna and fibula, very long metacarpals and longer phalanges on the forelimbs, which are attached to their wings in order to extend them and allow them to fly.
- There is an extended ridge of bone on the sternum, called a **keel**, to which the muscles responsible for flying are attached.
- They have developed an extra bone on the foot called the **calcar**, to support the skin that lies between the tail and the hindlimbs.

Hopping mammals

Hopping is an unusual method of movement but has evolved separately in a number of different mammals, the most common being rabbits. Hopping allows animals to move rapidly and quickly change direction, and is thought to have evolved to allow small mammals to avoid predators.

- The tibia and fibula in the hindlegs are relatively much longer. This, along with powerful hindlimb muscles, allows these mammals to hop.
- The forelimbs are much smaller than the hindlimbs.
- There is greater separation between the lower vertebrae, as well as an extended lumbar, to allow for more flexibility and extension when hopping. This gives the spine a curved shape but the drawback is that this curved spine is prone to breaking.
- The sacrum and the pelvis are fused to help withstand the forces caused by hopping. The ankles and toes also act like shock absorbers.
- The rabbit's skeleton is lighter than similarly sized mammals, but this does make it weaker and more prone to breaking, especially the tibia.
- Rabbits have four digits on the hindlimbs but five on their forelimbs.

Running mammals

Running animals, such as horses and cheetahs, are more formally known as cursors or cursorial animals. Such mammals tend to be relatively large.

Cheetahs are specially adapted for short but very high-speed sprints:

- They have longer metatarsals and metacarpals, giving them longer limbs and a greater stride length.
- They run on their toes and only semi-retract their claws to give them better grip while running.
- They have a small clavicle, not attached to the scapula, which allows for greater mobility and which in turn allows for a greater stride length.
- They have greater movement of the pelvis, for a quicker stride rate and to further increase stride length.
- They have a very flexible spine, to allow the whole body to curve and stetch with each stride. This allows them to create more power and to cover more ground with each stride.
- They have a long tail which acts like a rudder to help keep them balanced while running at speed, especially when cornering.
- They have a small and aerodynamic skull and ribcage.

Figure 29 A cheetah running at high speed – notice the curvature of the spine and a stride length which is so large that the forelegs and hindlegs cross over.

- They have a large nasal passage for more efficient breathing, needed for fast running. However these large nostrils leave little room for large teeth, so they have smaller teeth than other big cats. This affects their ability to fight.

Horses are adapted to run for long distances:

- They have fused their ulna and radius, as well as their tibia and fibula, so their limbs can take the impact of running given the horse's size and weight.
- The horse runs on a hoof which is actually a single toe. The horse evolved to have only one large and strong toe as their ancestors became bigger and heavier. A downside is that the remnants of the other much reduced toes can get damaged and cause inflammation.
- Horses have a large scapula, for muscle attachment at the top of the foreleg to provide strength, but only tendons and ligaments in the lower leg, to reduce weight and the energy used to run. The downside is that they are prone to damaging their ligaments and tendons, which take a long time to heal.
- Their limb bones are elongated to increase stride length, so they can cover more ground with each stride.
- They have a large rib cage to house their large lungs, which are needed for running longer distances.

Questions

1 State the names of the bones labelled in the diagram. (7)

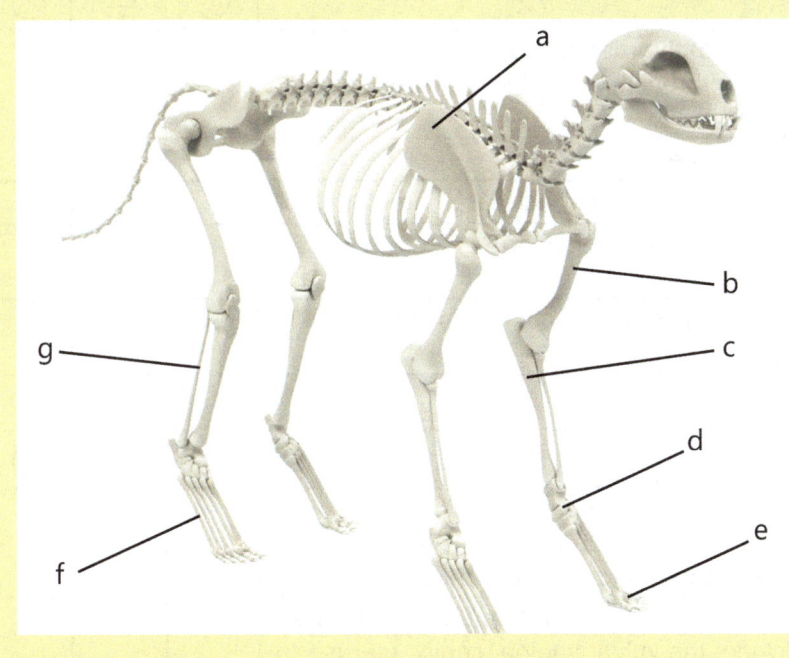

a _____
b _____
c _____
d _____
e _____
f _____
g _____

2 State three functions of the skeleton. (3)

3 What is the particular role of the axial skeleton? (2)

4 Explain the differences between tendons and ligaments. (2)

5 Discuss three advantages of the mammalian musculoskeletal system. (6)

6 Outline the main adaptations to the musculoskeletal system that are seen in horses. (4)

Learning Outcome 1 – Revision checklist

LO1 Understand the structure and function of biological systems in animals

1.1	☺	😐	☹
I can identify and describe the components of blood: plasma, erythrocytes, leukocytes and platelets.			
I can describe how blood cells form.			
I can identify and describe the components of the heart and their function: four chambers, aorta, vena cava, pulmonary vein, pulmonary artery, bicuspid and tricuspid valves, chordae tendinae, sino-atrial node, atrioventricular node, bundle of His and Purkynje fibres.			
I can explain the role each of these components plays in the heartbeat.			
I can label a diagram of the heart and double circulatory system.			
I can identify and describe the structure and function of blood vessels: capillaries, veins and arteries.			
I know about double circulatory systems and the animals that use them.			
I know about single circulatory systems and the animals that use them.			
I know about closed circulatory systems and the animals that use them.			
I know about open circulatory systems and the animals that use them.			

1.2	☺	😐	☹
I can describe the structure and function of the mammalian respiratory system: nasal chambers, larynx, trachea, bronchi, bronchioles, lungs, alveoli and diaphragm.			
I can label a diagram of the respiratory system.			
I know the respiratory system has adapted in some species of mammal.			
I understand how ventilation of the lungs and gas exchange in the alveoli takes place.			
I know about comparative adaptations of the respiratory system in fish, amphibians, birds and invertebrates.			

1.3	☺	😐	☹
I can describe the location, structure and function of the male reproductive systems to include: the penis, prepuce, urethra, bulbus glandus, epididymis, vas deferens, testes (testosterone, oestrogen) and prostate gland.			
I can describe the location, structure and function of the female reproductive systems in viviparous animals to include: the vulva, vagina, cervix, uterus, oviduct and ovaries.			
I can label diagrams of the male and female reproductive systems.			
I understand comparative adaptations of the reproductive system in a range of animals including cats, whales and pigs.			
I understand the stages of the oestrus cycle including the roles of the hormones oestrogen, progesterone, LH and FSH.			
I understand human influence on breeding, such as hormonal injections in bitches and broodmares.			

	☺	😐	☹
I understand the stages of the sexual reproduction cycle to include: copulation, fertilisation, implantation, gestation and parturition.			
I understand the term viviparous.			
I understand the terms oviparous, ovoviviparous, egg-laying mammals and marsupials, and understand the adaptations to the reproductive cycle in these animals.			

1.4	☺	😐	☹
I can identify the structure and function of the excretory system: kidneys, ureters, bladder and urethra.			
I can label diagrams of the excretory system.			
I understand the terms ultrafiltration and reabsorption and how this relates to kidney function.			
I understand the differences between voluntary and involuntary control of the bladder.			
I understand comparative adaptations to the excretory system in birds, desert mammals and aquatic animals.			

1.5	☺	😐	☹
I can identify the structure and function of the mammalian musculoskeletal system.			
I can label diagrams of the mammalian musculoskeletal system.			
I understand the advantages and disadvantages of evolutionary adaptations to the mammalian musculoskeletal system.			
I can understand the importance of the musculoskeletal system.			
I understand adaptations to the skeleton of: aquatic animals, flying mammals, hopping mammals, running mammals.			

Learning Outcome 1 - My Revision Goals

Consider each of the points in the revision checklist. For each point, what do you still need to do to make sure you are fully prepared for your exam?

On a separate piece of paper, or using the space below, write down each of the tasks you identify. (If this book was given to you by your tutor then ask before writing in it!)

LO1 Understand the structure and function of biological systems in animals

LO2 Understand control mechanisms in animals

2.1 Structure and function of hormonal mechanisms in the endocrine system

> **In this topic you will learn about:**
>
> - The structure and function of the endocrine system, including the location of the major endocrine glands.
> - The role of cell receptors within the endocrine system.
> - Homeostasis of water and blood sugar levels.
> - The difference between circulating hormones and local hormones.

The **endocrine system** uses **hormones** to communicate with and control organs in the body. Hormones are released into the bloodstream by **endocrine glands**.

Hormones act like chemical messengers that travel from the glands, through the blood, to different organs.

- Some cells within organs have **receptors** on their surface.
- Receptors are proteins that bind to specific hormones.
- You can think of receptors as being like a lock and hormones like a key – binding only happens when the correct key (hormone) and lock (receptor) combine.
- The hormone/receptor binding process allows the hormones to change the behaviour of other cells within that organ.

In this way, the endocrine system allows the behaviour of organs to be controlled remotely.

The major endocrine glands are as follows:

Hypothalamus

The **hypothalamus** is a small part of the brain that constantly monitors a number of the body's parameters – for instance temperature, heart rate, blood pressure and water concentration in the blood. It detects changes in these parameters and releases hormones that stimulate or supress the pituitary gland. In this way, the hypothalamus controls the pituitary gland. The hypothalamus also produces anti-diuretic hormone (ADH) and oxytocin, which are then stored and released by the pituitary gland. (See section 1.4 for more on the role of ADH in the exretory system).

Pituitary Gland

This is often called the 'master gland' because it releases hormones that stimulate other glands in the body – the adrenal, thyroid, ovaries and testes. Thus the **pituitary gland** controls these other glands. There are two parts to the pituitary gland: the anterior (front) and posterior (rear).

The anterior pituitary gland produces hormones including:

- **Adrenocorticotropic hormone** – this stimulates the adrenal gland.
- **Thyroid-stimulating hormone** – this stimulates the thyroid gland.
- **FSH (follicle stimulating hormone)** – this stimulates follicles within ovaries to grow and mature.
- **LH (luteinising hormone)** – this also acts

on the ovaries and stimulates ovulation.

- **Prolactin** – this hormone is associated with milk production in mammals.
- **Growth hormone** – this regulates the repair of tissue and promotes growth in young animals.

The posterior pituitary gland stores and releases the following:

- **Anti-diuretic hormone (ADH)** – this does not stimulate another gland but instead directly acts on the kidneys and blood vessels, controlling the amount of water content in the blood.
- **Oxytocin** – a hormone which acts on the smooth muscle of the uterus, causing contractions during parturition. It also has a role in social bonding.

> **Link**
> For more on the role of FSH and LH within the oestrous cycle, and the role of prolactin and oxytocin in parturition, see section 1.3

Pancreas

This organ detects and regulates the amount of glucose (a type of sugar) in the bloodstream.

All cells in the body require glucose to function. The body uses glucose as a fuel and obtains it from breaking down carbohydrates when digesting food. This means that blood glucose levels constantly change but must be present at all times. It is dangerous for the animal if blood glucose levels are too high (hyperglycaemic) or too low (hypoglycaemic).

The **pancreas** detects and controls the levels of glucose in the blood using the **islets of Langerhans**. There are two types of cell in the islets of Langerhans: alpha and beta. The alpha cells produce **glucagon**, and the beta cells produce **insulin**. These hormones keep the blood glucose balanced and maintained at the correct level.

- Glucagon causes the stored glycogen in the liver to be released into the blood. The production of glucagon increases the level of blood sugar in the blood.
- Insulin allows glucose in the blood to either be absorbed by cells in the body or to be converted to glycogen and stored in the liver. In either case, the production of insulin lowers blood sugar levels in the blood.

There are two types of diabetes, both of which affect the body's ability to regulate the blood glucose:

- Type 1: The animal has a problem producing insulin. The animal requires injections of insulin to regulate their blood glucose and stop it from getting too high and causing hyperglycaemia.
- Type 2: The animal has become insulin resistant. This type can be controlled by diet.

Adrenal

The **adrenal glands** are located on top of the kidneys and produce a number of different hormones. There are two parts to the adrenal gland: the outer cortex and inner medulla.

The medulla produces:

- **Adrenaline** – also known as epinephrine, this is a fast-acting hormone that acts on most tissues in the body and is responsible for a 'fight or flight' response to a stressful situation. Amongst other things it causes the heart and breathing rate to increase, providing more blood to the muscles and brain for their immediate use.
- **Noradrenaline** - also known as norepinephrine, this is also a fast-acting hormone that has a role in the 'fight or flight' response. It helps increase or maintain blood pressure by constricting blood vessels. It also increases alertness and attention.

> **Link**
> For more on 'fight or flight' see the autonomic nervous system in section 3.2.

The cortex produces:

- **Cortisol** – the so-called 'stress hormone' that gives the body an energy boost in times of stress by controlling the way in which the body processes nutrients.
- **Aldosterone** – this hormone instructs the

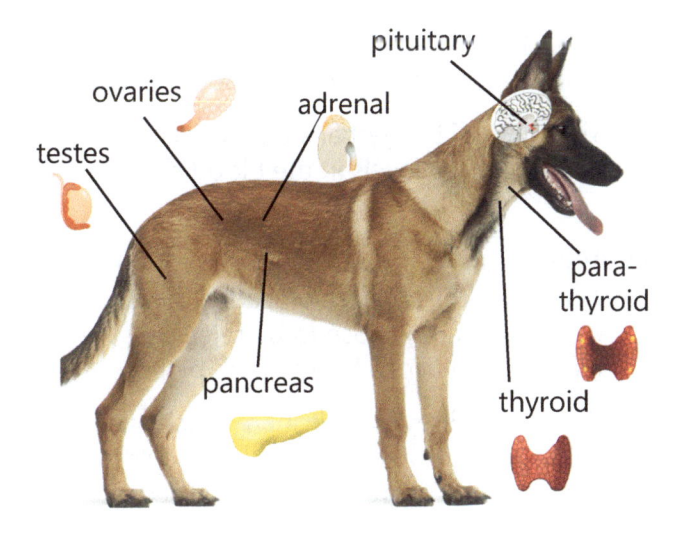

Figure 30 The endocrine system

kidney to remove or retain certain salts in the body. It helps regulate the balance of water and salts in the bloodstream.

> **Link**
> For more on the function of the kidney see section 1.4.

Thyroid

The **thyroid gland** is responsible for metabolism. Metabolism refers to all the chemical processes that take place in the body to support life. It determines how quickly energy is used by an animal.

The thyroid gland secretes two hormones: triiodothyronine (**T3**) and thyroxine (**T4**). They both cause the body's cells to work faster. The hypothalamus tracks the levels of these hormones in the blood and instructs the pituitary gland to produce more or less thyroid-stimulating hormone. Thyroid-stimulating hormone causes the thyroid to produce more triiodothyronine (T3) and thyroxine (T4).

There are two conditions of the thyroid. The first is overactive thyroid (hyperthyroidism) which causes the animal to be very active, speeding up their heart rate and increasing appetite but is associated with weight loss and decreased sleep. The second is underactive thyroid (hypothyroidism) which decreases the animal's activity levels and slows the heart rate, leading to weight gain, joint pain and increased levels of sleep.

The thyroid also produces another hormone called **calcitonin** which helps maintain calcium levels along with para-thyroid hormone.

Para-thyroid

The **para-thyroid gland** regulates the levels of calcium in the body. It does this by producing a hormone called **parathyroid hormone**. The correct levels of calcium are important for the proper functioning of the nervous system and in order to keep bones strong.

- Increased levels of parathyroid hormone increase the rate of bone being broken down as part of the normal process of cell maintenance and repair. An increase in bone breakdown leads to more calcium in the blood, as a major component of bone is calcium.
- The calcitonin produced by the thyroid gland has the opposite effect. Increased levels of calcitonin reduce the rate of bone breakdown, and therefore the amount of calcium being released into the blood.

This is another example of homeostasis.

Ovaries

The ovaries are part of the female reproductive system and produce the female sex hormones **oestrogen** and **progesterone**, which regulate the female reproductive system. Oestrogen has roles in female secondary sex characteristics and also thickens the lining of the uterus ready for copulation to occur. Progesterone maintains the uterus lining, especially during gestation.

> **Link**
> See section 1.3 for more on the ovaries.

Testes

The testes are part of the male reproductive system and produce the male sex hormone **testosterone**. Testosterone is responsible for the development of male sex characteristics and stimulates the production of sperm.

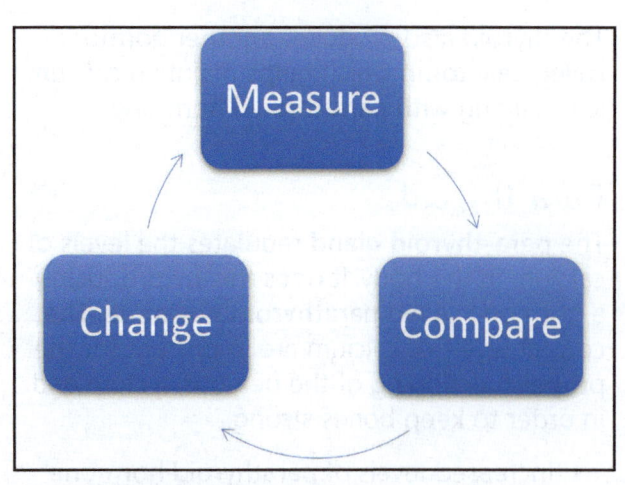

Figure 31 Negative feedback process

Homeostasis

Homeostasis is the process by which bodily conditions are maintained within the correct range. For instance, most mammals must keep their body temperature within a narrow range (typically around 36°C- 40°C) at all times or risk death.

There are two ways to regulate or control the body: positive and negative feedback.

Positive feedback amplifies the response and intensifies it. Examples include:

- Partruition, when pressure on the cervix releases oxytocin, which causes contractions, leading to more pressure on the cervix and the release of more oxytocin.
- Blood clotting, where platelets at the injury site release chemicals to attract more platelets, which in turn release more chemicals. This speeds up clotting until the clot is large enough.

Negative feedback loops are key for homeostasis. They consist of the following three stages:

- Measure
- Compare
- Change

An example of homeostasis is the control of blood glucose levels. It works as follows:

Measure

- Blood glucose levels are monitored by the islets of Langerhans in the pancreas.

Compare

- The islets of Langerhans alpha cells detect when glucose levels in the blood are low.
- The islets of Langerhans beta cells detect when glucose levels in the blood are high.

Change

- The alpha cells release glucagon, which causes stored glucose to be released, increasing glucose levels in the blood.
- The beta cells release insulin, which causes glucose to be absorbed by the body's cells and the liver, decreasing blood glucose levels.

It is called negative feedback because the process always tries to cancel out any changes, in order to keep it within acceptable levels. For instance, when glucose levels are too high insulin is released which acts to negate the high levels of glucose.

Another example of homeostasis is the control of water levels in the body. Water is essential for all bodily processes, so it is crucial to keep water levels above a certain level. The process works as follows:

Measure

- The hypothalamus detects the water levels in the blood.

Compare

- If water levels in the blood get too low, the hypothalamus sends a signal to the pituitary gland.

Change

In response the pituitary gland releases anti-diuretic hormone (ADH). The lower the amount of water in the blood, the more ADH is released.

ADH travels to the kidneys and opens the pores in the collecting duct of the nephron. This allows more water to be reabsorbed back into the bloodstream and less water to be expelled in urine. This conserves more water in the body.

When detecting low levels of water, the

hypothalamus also sends signals to other parts of the brain which are interpreted as thirst, causing the animal to drink more. The hypothalamus continually measures water levels. As they increase, the production of ADH decreases, so the kidneys can pass more water out in urine.

> **Link**
> See section 1.4 for more on the role of the kidneys in the excretory system.

Circulating hormones and local hormones

There are two types of hormones:

- **Circulating hormones** flow through the bloodstream to different parts of the body. The cells that need to react to the hormone have specific receptors on their surface that the hormones bind to, triggering a response. Other cells do not have the receptor for this hormone. All of the hormones discussed earlier are examples of circulating hormones.

- **Local hormones** are not released into the bloodstream and therefore do not circulate through the body. Instead, they act on cells very near to where they are released.

One example of a local hormone is histamine, which is released by mast cells as part of the body's immune response, and causes localised inflammation. This inflammation can be reduced by the medications called antihistamines.

Another example of local hormones are prostaglandins. Their function varies depending on where in the body they are made. For example there are prostaglandins in seminal fluid, which aid the movement of sperm through the female's reproductive tract and trigger contractions of smooth muscle in the uterus, to move the sperm closer to the egg.

All the glands discussed here are called endocrine glands which secrete hormones. Another set of glands are called exocrine glands, which secrete substances onto the surface of the body or organs via ducts. Examples of exocrine glands include salivary glands and sweat glands. The pancreas also has exocrine functions, such as secreting pancreatic juices (containing enzymes) to the duodenum, in order to break down carbohydrates, proteins and fats.

Questions

1. Label the endocrine glands organs of this female dog. (6)

2. Describe main stages in the homeostasis of blood sugar levels. (3)

3. Name four hormones released by the pituitary gland. (4)

4. Explain the role of hormones and how they function. (3)

5. Discuss the role of the adrenal gland. (3)

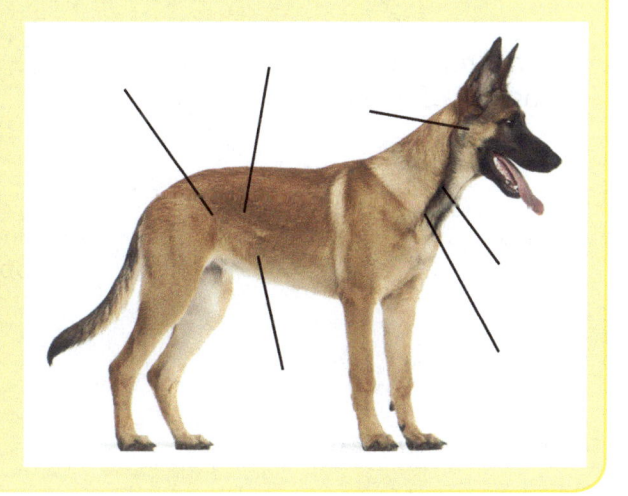

2.2 Structure and function of the lymphatic system

In this topic you will learn about:

- The structure of the lymphatic system, including nodes (glands) and vessels.
- The function of the lymphatic system.
- The role of T-lymphocytes and B-lymphocytes within the immune system.

Structure of glands and vessels

The lymphatic system is part of the immune system. It circulates a colourless liquid called **lymph** through a network of lymph vessels and lymph nodes (also known as lymph glands).

As blood delivers nutrients and oxygen to all cells in the body, most of it returns to veins in the circulatory system. However, a small amount of fluid enters the surrounding tissues instead, and drains into the lymphatic system via lymph capillaries. Once inside the lymphatic system this fluid is called lymph.

- **Lymph nodes** (also known as lymph glands) are small organs that are distributed throughout the body. There are typically hundreds of lymph nodes. They act as filters which trap dead cells, cancerous cells or pathogens. They contain **lymphocytes** which are part of the immune system and which can destroy any detected pathogens.

- Lymph nodes are connected by a series of **lymph vessels**. Lymph vessels transport lymph towards the circulatory system and away from body tissues. They are thin tubes, similar to veins, with valves to ensure that lymph can only flow in one direction. The smallest lymph vessels are called lymph capillaries. Lymph capillaries

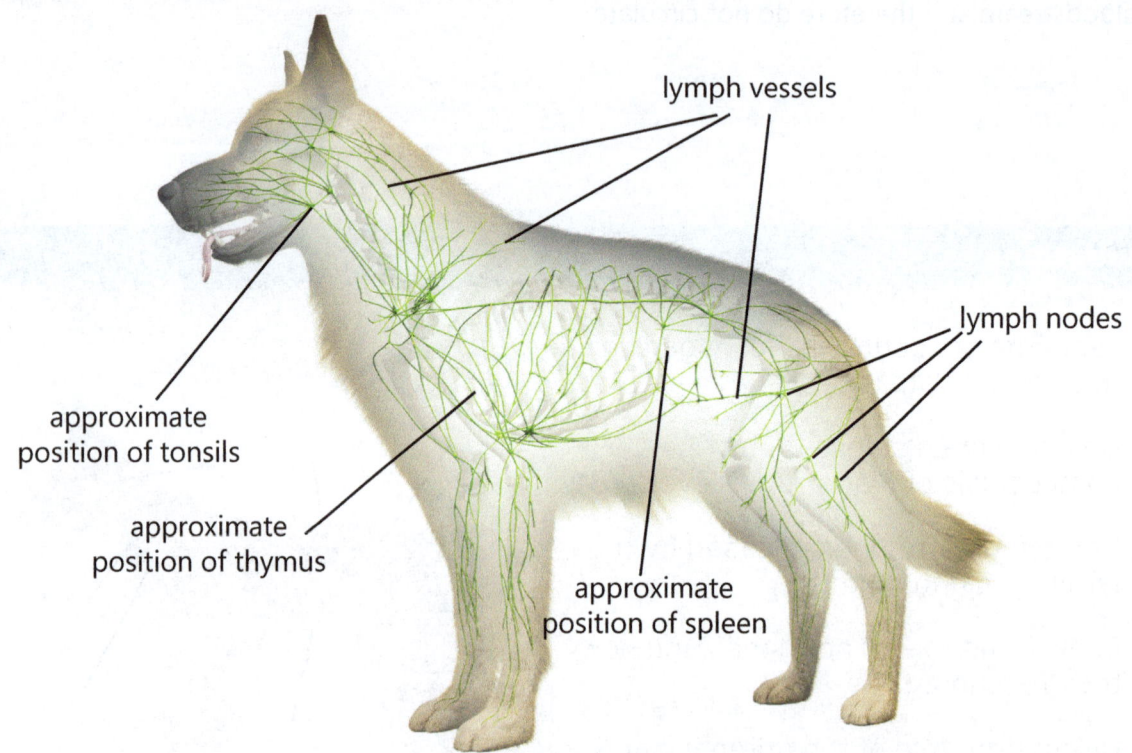

Figure 32 The lymphatic system of a dog

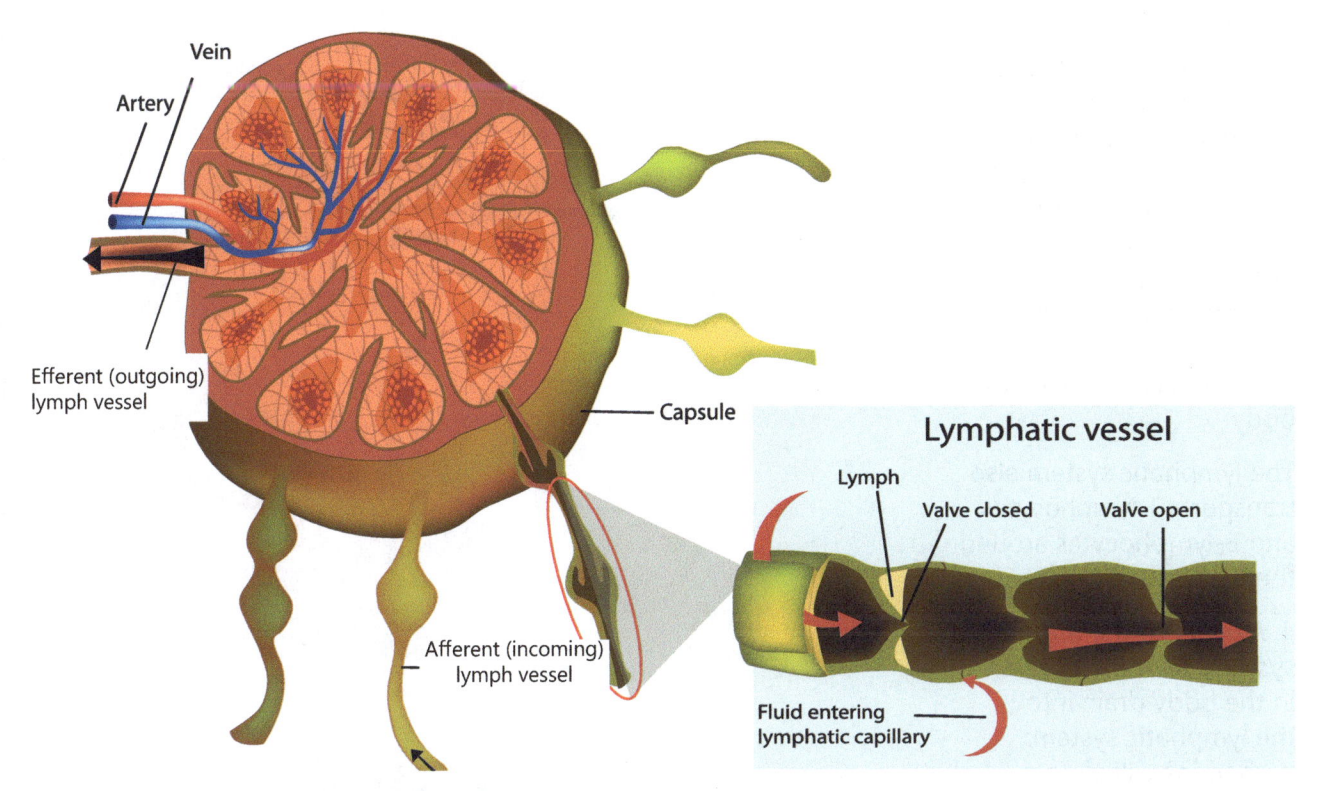

Figure 33 A lymph node (lymph gland) and a lymph vessel

collect fluid from tissues and their walls are only one cell thick.

- Lymph ducts return lymph to the circulatory system.

The lymphatic system does not have an equivalent organ to the heart. This means lymph is not pumped around the body but instead either relies on gravity or is squeezed around the system using muscles. This means that lymph circulation is much slower than blood circulation.

The lymphatic system is not a closed, self-contained loop. Instead, lymph is collected from all parts of the body and returned to circulatory system.

There are several other organs that also play a role in the lymphatic system:

- The spleen – located next to the stomach the spleen stores blood cells, filters out old blood cells and breaks down old erythrocytes. It also controls the amount of erythrocytes and leukocytes in the body.
- The thymus – T-lymphocytes mature here (see next section on the immune system). The thymus is important for young animals.
- The tonsils – located in the throat, these are major lymph nodes which act as filters, destroying any pathogens entering the body through the mouth.

Functions of the lymphatic system

The lymphatic system has four key functions:

1. **To drain excess fluid from tissue.** As some blood is continually lost to the surrounding tissues, this fluid needs somewhere to go to. Without a lymphatic system, the tissues in the body would painfully swell up with fluid and become damaged.

2. **Absorption of fats.** Whilst carbohydrates and proteins are broken down and absorbed into the bloodstream in the small intestine, longer fatty acid chains are absorbed by the lacteals which are connected to the lymphatic system. Lacteals are found in the villi in the intestines - see Unit 304, section 1.3.

3. **Transport of materials.** Lymph transports water, cells, immune responses, fat and waste material around the body.

Waste products from cells pass into the surrounding tissues and accumulate in the lymphatic system. Fat also collects in the lymphatic system, along with any pathogens that the immune system has disabled. Lymph circulates through the lymphatic system until it eventually empties back into the bloodstream. Waste in the blood can then be excreted from the body

The lymphatic system also transports T-lymphocytes and B-lymphocytes around the body.

4. **As part of the immune system.** As most tissues in the body drain into the lymphatic system, pathogens will almost always make their way into it. Within the lymphatic system are components of the adaptive immune system. This is made of two types of leukocytes (white blood cells): **B-lymphocytes** and **T-lymphocytes**.

B-lymphocytes (also known as **B-cells**) are formed and mature in bone marrow. (The **B** is for **B**one marrow). They gather in the lymph nodes. B-lymphocytes recognise pathogens and then manufacture the correct antibody in response. They can only to respond to pathogens in the blood and lymph.

T-lymphocytes (or **T-cells**) are formed in the bone marrow but mature in the thymus. (The **T** is for **T**hymus). The thymus is located in the chest. **T-lymphocytes** play a similar role to B-lymphocytes but they can locate pathogens that have invaded cells of the body.

Lymph nodes act as gateways where B-cells and T-cells first recognise and react to pathogens. Part of that reaction is the rapid production of cells with the correct antibodies. This is why lymph nodes become swollen when animals are ill. The location of swollen lymph nodes helps veterinary professionals to locate, track and identify infections.

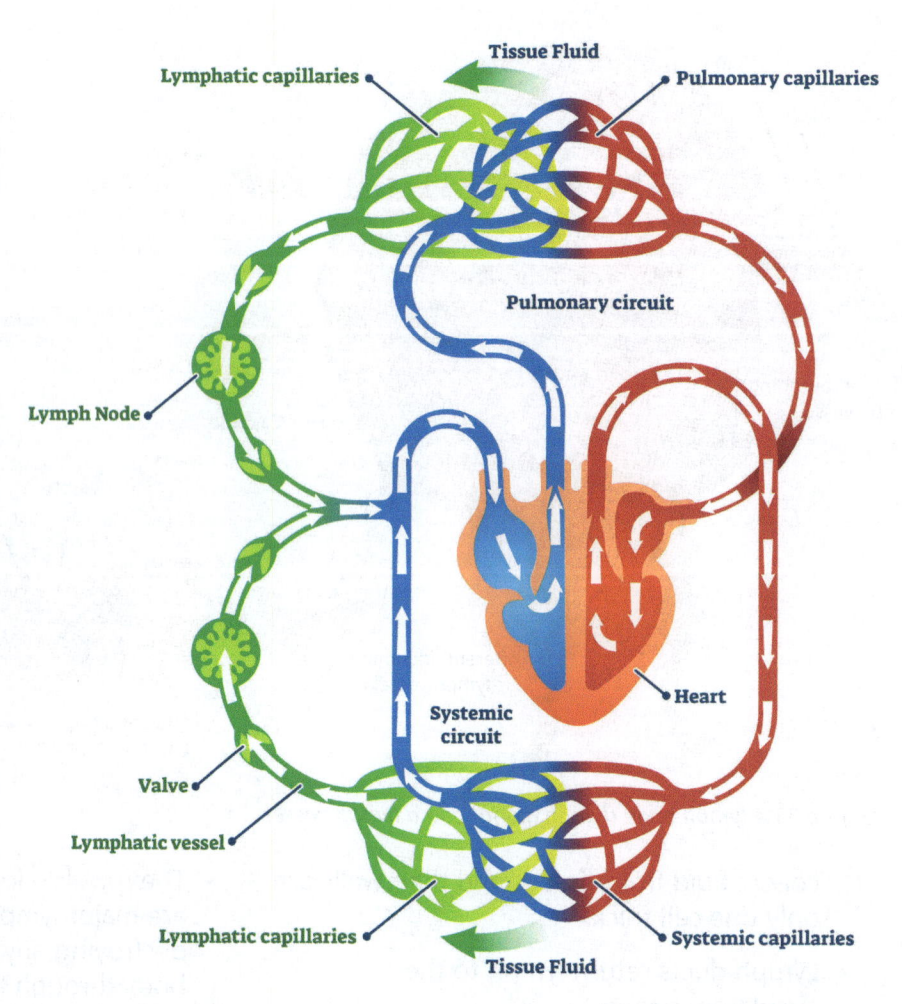

Figure 34 A schematic diagram showing the interaction of the lymphatic system with the blood circulatory system. Note that in reality the lymphatic system is distributed on all sides of the body.

Link
For more on leukocytes see section 1.1.

For more on the excretory system see section 1.4.

More detail on B-lymphocytes and T-lymphocytes

Non-lymphocyte white blood cells, discussed in section 1.1, are part of the innate immune system. This system acts quickly when pathogens are detected but cannot always disable the pathogen because their response is generic. B-cells and T-cells make up the adaptive immune system. It learns how to respond to different pathogens, and then remembers how it did so, in case the pathogen reappears in the future.

All cells have molecules on their surface. It is these cells which the immune system recognises and identifies. They are called antigens, short for 'antibody generators'. Each pathogen has a unique set of antigens, and it is the antigens which trigger an immune response. (The body's own cells also have these molecules on their surface, but are called self-antigens, which the immune system ignores.)

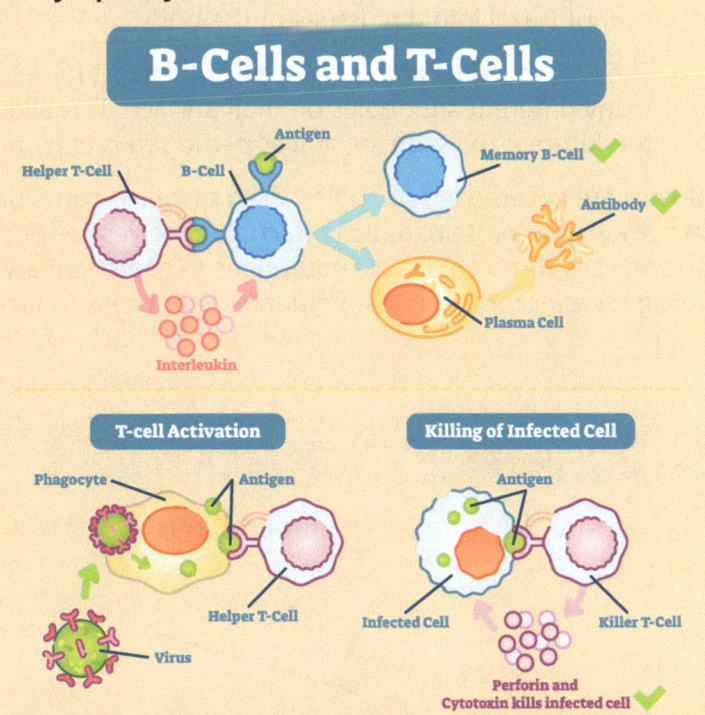

Figure 35 The activation of T-cells and B-cells. An antigen-presenting T-cell helps to activate a B-cell with the help of interleukin, a type of cytokine. Once activated, memory B-cells and plasma cells are produced in large numbers. The T-cell itself is presented with an antigen, is activated, and goes on to produce more helper T-cells and cytotoxic (killer) T-cells. These kill infected cells using chemicals which are deadly to animal cells, called cytotoxins.

As B-cells and T-cells are mainly located in lymph nodes, they need to be alerted to the presence of a pathogen elsewhere in the body. This is done by one of the two versions of a monocyte, called a dendritic cell. A dendritic cell picks up an antigen from a pathogen and takes it to the lymph nodes, where T-cells inspect the antigen to see if they recognise it. If they do then they become activated. Once activated, T-cells replicate and differentiate into two different types of T-cell.

- Helper T-cells are able to continue to detect the presence of antigens brought to them, through their T-cell receptors. This causes the helper T-cells to quickly multiply in number. They produce a range of chemical messengers called cytokines, which perform a number of roles.
- Cytotoxic T-cells are activated by helper T-cell cytokines. Cytotoxic T-cells kill pathogens that have already infected the body's cells by killing the cells themselves. They do this by releasing chemicals that destroy cell membranes, and hence they are able to destroy pathogens within the body's tissues.

Helper T-cells also travel to the site of the pathogen and release a second type of cytokine, this time to alert the innate immune system to generate cells to perform phagocytosis.

Helper T-cells also produce a third type of cytokine that helps activate B-cells. Once activated, B-cells differentiate into two different types of B-cell which quickly multiply in number:

- They can become plasma cells, which multiply in number and generate large amounts of the correct antibody which are then released them into the body, to fight the specific

Continued...

pathogen. Antibodies can travel around the lymphatic system and bloodstream but cannot travel into the tissues of the body.

- They can become memory B-cells, which last for a lot longer than plasma cells, and retain many different antibodies on their surface. This allows the B-cells to be activated again in the future if the same or similar pathogen is encountered.

B-cells are covered in antibodies, each of which corresponds to a specific antigen. Antibodies are 'Y'-shaped proteins which attach themselves to specifc antigens. By attaching themselves to antigens they can disable pathogens (e.g. prevent a virus from being able to replicate), or make it easier for the innate immune system cells to recognise which cells to destroy.

Questions

1. Identify the structures labelled in the diagram below. (2)

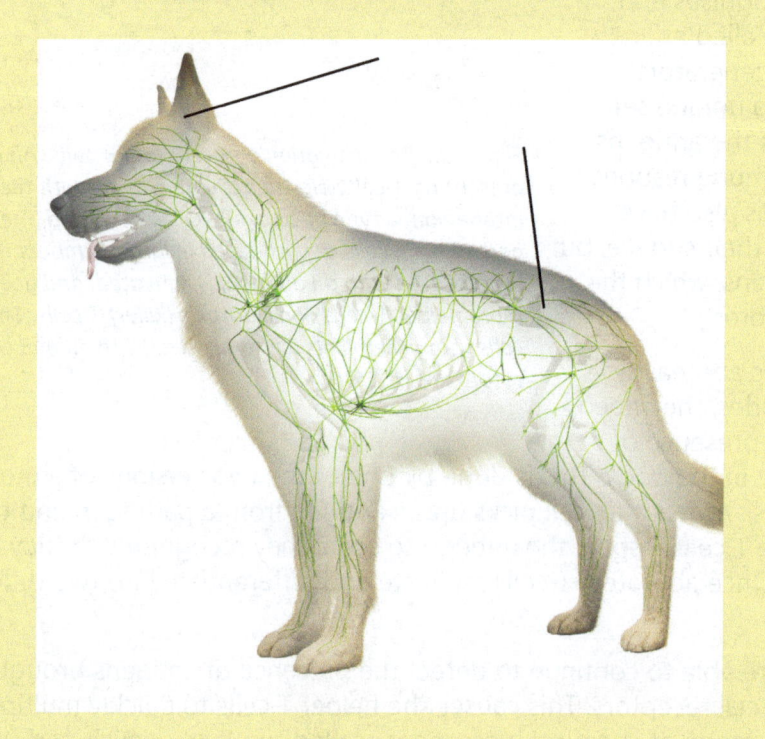

2. State two functions of the lymphatic system and describe how it carries out each of these functions. (4)

3. Compare and contrast the lymphatic system and the circulatory system. (4)

4. Outline the similarities and differences between T-lymphocytes and B-lymphocytes. (4)

2.3 Structure, function and adaptations of the thermoregulatory system

In this topic you will learn about:

- Ectotherms.
- Endotherms.
- Homeostasis of body temperature.

An animal's body needs to be kept within a narrow temperature range to be able to function correctly. If it gets too hot (hyperthermia) or too cold (hypothermia) then the body stops functioning properly, which can be fatal. It is very important that animals can regulate their body temperature.

Thermoregulation refers to the methods animals use to keep their bodies within the correct temperature range. Animals can either be:

- Ectotherms – **ecto** means external/outside, and **therm** means heat.
- Endotherms – **endo** means internal/inside, and **therm** means heat.

Ectotherms

Ectotherms are animals that rely on their environment to maintain their body temperature. This is because their internal metabolic processes, which derive from eating and digesting food, do not generate enough heat to warm them. Instead, they alter their behaviour. For instance:

- If they are getting cold then they seek out warm places – such as finding somewhere to bask in the sun, or locating some warm rocks. Captive animals sit under a heat lamp.
- If they are getting too hot they will try and escape from the source of heat, burrow into the ground, find shade or move into water.

An ectotherm that is too cold becomes slow and sluggish. This is because the chemical processes that allow their muscles to move have slowed down.

- Because the environment is so important for their internal body temperature,

Figure 36 In captivity ectotherms need to be kept in a warm environment – such as this iguana basking under an infrared lamp

ectotherms often live in places that are warm and without too much variation in temperature.

- As ectotherms do not rely on food to keep them warm, they do not need to eat as regularly as endotherms.

- As ectotherms have a lower metabolic rate they must warm up before carrying out energy-intensive tasks.

- Ectotherms are commonly known as 'cold blooded' animals. However, this is a misleading term as ectotherms normally need to be warm in order to function normally.

- All reptiles, amphibians, insects and most fish are ectotherms.

Endotherms

Endotherms are animals that maintain their body temperature using internal metabolic processes. These processes require a constant supply of energy, which is provided by food. Endotherms must eat regularly to meet these metabolic demands.

Because endotherms maintain their temperature using internal metabolic processes, they can live in environments with a wider range of temperatures than ectotherms – providing there is enough food.

All mammals and birds are endotherms.

Homeostasis of body temperature

Most ectotherms have a wider range of acceptable body temperatures than endotherms. So, whilst a reptile might be sluggish if it gets too cold, they are able to survive this lowering of body temperature.

Endotherms, on the other hand, need their body temperature to stay within a fairly small range. For many species it is fatal it falls below or above this range. Endotherms have therefore developed homeostasis processes, using negative feedback, to ensure that body temperature remains within an acceptable range:

Measure

- The hypothalamus contains receptors that are able to monitor blood temperature, and also receives information from temperature receptors on the skin.

Compare

- Hypothalamus recognises the signals from these receptors and reacts when temperature is too high or too low.

Change: temperature too high

- The animal increases respiration rate (for example, by panting) to get rid of excess heat through evaporation of moisture on the tongue and in the mouth.

- Vasodilation occurs - this is where blood vessels in the skin widen and get closer to the skin surface, giving a red appearance. The skin surface is in contact with cooler air, allowing warm blood to radiate heat into the environment, which cools the body.

- Some animals have the ability to sweat but not all animals can sweat all over their body. For instance, a dog can only sweat a small amount from its paws. Sweat glands release sweat on the surface of the skin, and as the sweat evaporates it removes heat from the skin surface, cooling the animal down.

Change: temperature too low

- The body begins to shiver, caused by the skeletal muscles being instructed to vibrate. Muscle movements generate heat and warm the animal up. (This is why we feel warm after exercise).

- Vasoconstriction occurs - this is where blood vessels near the skin surface become narrower, reducing the amount of warm blood reaching the skin surface. This slows down radiation of heat from the skin into the cooler air. This ensures that internal organs are kept as warm as possible.

- Any hair or feathers on the animal stand on end, trapping warmer air next to the skin and keeping the colder air away from the skin.

The hypothalamus also signals to the rest of the brain if it detects that the animal is too warm or too cold, and the animal changes its behaviour accorditngly – for instance to find shade or warmth.

Link
See section 2.1 for more on homeostasis.

Did you know?
Some endotherms have also developed methods like hibernation to cope with extended periods of low temperatures or lack of food.

Questions

1. Define the term 'ectotherm' and give an example of one. (2)
2. What environments are ectotherms less suited to? (1)
3. List two advantages that an endotherm has over an ectotherm. (2)
4. Outline how homeostasis of body temperature functions in a dog. (3)

Learning Outcome 2 – Revision checklist

2.1	☺	😐	☹
I can locate the position of the following endocrine glands, and describe their function: hypothalamus, pituitary gland, pancreas, adrenal gland, thyroid gland, para-thyroid gland, ovaries, testes.			
I can explain the process of homeostasis of sugar levels in the blood.			
I can explain the role of cell surface receptors in the endocrine system.			
I know the difference between circulating hormones and locally acting hormones.			

2.2	☺	😐	☹
I can identify the structure and function of the lymphatic system, including lymph glands (nodes), lymph vessels and lymph ducts.			
I can label a diagram of the lymphatic system.			
I can describe the four functions of the lymphatic system.			
I can describe the role of the thymus, T-lymphocytes and B-lymphocytes in the immune system.			

2.3	☺	😐	☹
I can describe the thermoregulatory system of an endotherm.			
I can describe the thermoregulatory system of an ectotherm.			
I can describe homeostasis in relation to body temperature.			

Learning Outcome 2 - My Revision Goals

Consider each of the points in the revision checklist. For each point, what do you still need to do to make sure you are fully prepared for your exam?

On a separate piece of paper, or using the space below, write down each of the tasks you identify. (If this book was given to you by your tutor then ask before writing in it!)

LO3 Understand the neural control mechanisms in animals

3.1 Gross anatomy of the brain

3.2 Neural control mechanisms in animals

> **In this topic you will learn about:**
> - The function and anatomy of the hindbrain.
> - The function and anatomy of the midbrain
> - The function and anatomy of the forebrain.
> - The Central Nervous System.
> - The Peripheral Nervous System, including the afferent and efferent pathways.
> - The Autonomic Nervous System, including the sympathetic and parasympathetic nervous system.

At a very early stage of embryo development, three distinct bulges appear in the neural tube. These eventually form distinct regions of the brain known as:

- the **forebrain**
- the **midbrain**
- the **hindbrain**

These three regions of the brain represent hundreds of millions of years of evolution. In evolutionary terms the hindbrain is the oldest part of the brain – this means it is present in more primitive lifeforms from many millions of years ago. As life became more complex, and animals more sophisticated, the brain continued to evolve and developed what became a midbrain and forebrain. The hindbrain controls more primitive responses whilst the forebrain is where higher-level problem-solving and reasoning takes place.

Hindbrain

The hindbrain is at the top of the spinal cord, and connects it to the rest of the brain.

Medulla

The **medulla oblongata** is at the base of the brain stem between the brain and spinal cord. It is responsible for maintaining autonomic functions such as:

- heart rate
- breathing rate
- blood pressure
- reflexes such as swallowing and coughing.

> **Link**
> For more on the autonomic system see section 3.2 The Autonomic Nervous System.

Cerebellum

The cerebellum is at the base of the brain and is deeply folded to increase surface area. It is responsible for coordination, balance and movement, allowing limbs to move in a co-ordinated way. The cerebellum allows animals to maintain balance without having to consciously think about it.

Pons

Pons means 'bridge' in Latin, and this area of the brain is exactly that – a bridge between the medulla oblongata and the midbrain along which signals travel.

Midbrain

The midbrain sits above the pons and is at the top of the brain stem.

Reticular Formation

The **reticular formation** (RF) is a complex network of neurons. 'Reticular' comes from a Latin word meaning 'net'

Its role is to receive and transmit information between the brain and the body. There are two types of information:

- sensory information e.g. hearing and sight
- motor information – the position and movement of muscles in the body

The midbrain has a role in sleep, consciousness and arousal states. In order to sleep, the flow of sensory information through the reticular formation needs to slow down. This is why animals often choose quiet places to sleep in. Less sensory information allows less motor information to pass through too. This allows muscles in the body to relax, including those in the eyelids.

Low levels of activity in the reticular formation are associated with sleep, unconsciousness and comas.

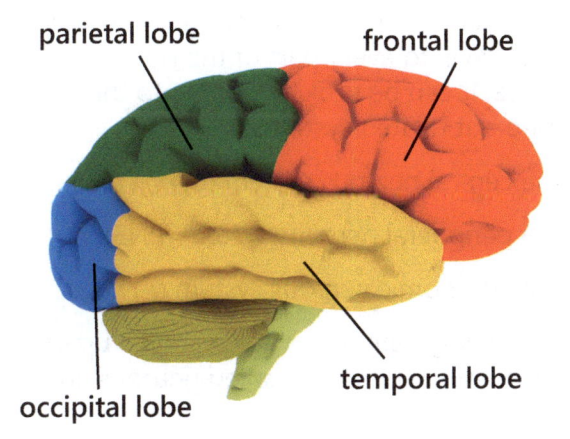

Figure 37a The lobes of the cerebral cortex

Forebrain

Cerebral cortex

This the large outer layer of the brain where higher-level processing takes place. It is made of two layers: the grey and white matter. In most mammals it has deep folds to increase surface area and animals with more folds are generally more intelligent. The folds also allow nutrients to reach deeper parts of the brain.

The cerebral cortex has general functions of higher-level thinking and voluntary thoughts. Any behaviour that might be described as 'intelligent' takes place in the **cerebral cortex**. As it is so large it gets split into four lobes:

- Frontal lobe: for planning and decision-making.
- Temporal lobe: for auditory information processing and memory.
- Parietal lobe: for sensory information processing, including touch, navigation and spatial processing.
- Occipital lobe: for visual processing

> **Link**
> See section 2.1 for more information on the endocrine system.

Limbic system

The **limbic system** is made up of several different structures in the brain. Together, they are responsible for emotions – such as fear, aggression and pleasure – and an animal's response to those feelings. The most important parts of the brain that contribute to the limbic system are:

- The **thalamus**: The main function of the thalamus is as a 'relay station', responsible for sending sensory information from anywhere in the body to the cerebral cortex, where it is processed and interpreted. Sensory information is not just limited to organs such as the eyes – it can be, for instance, the sensation of aching muscles, or the feeling of warmth on the face. All emotions are a response to some sensory information which is transmitted by the thalamus.

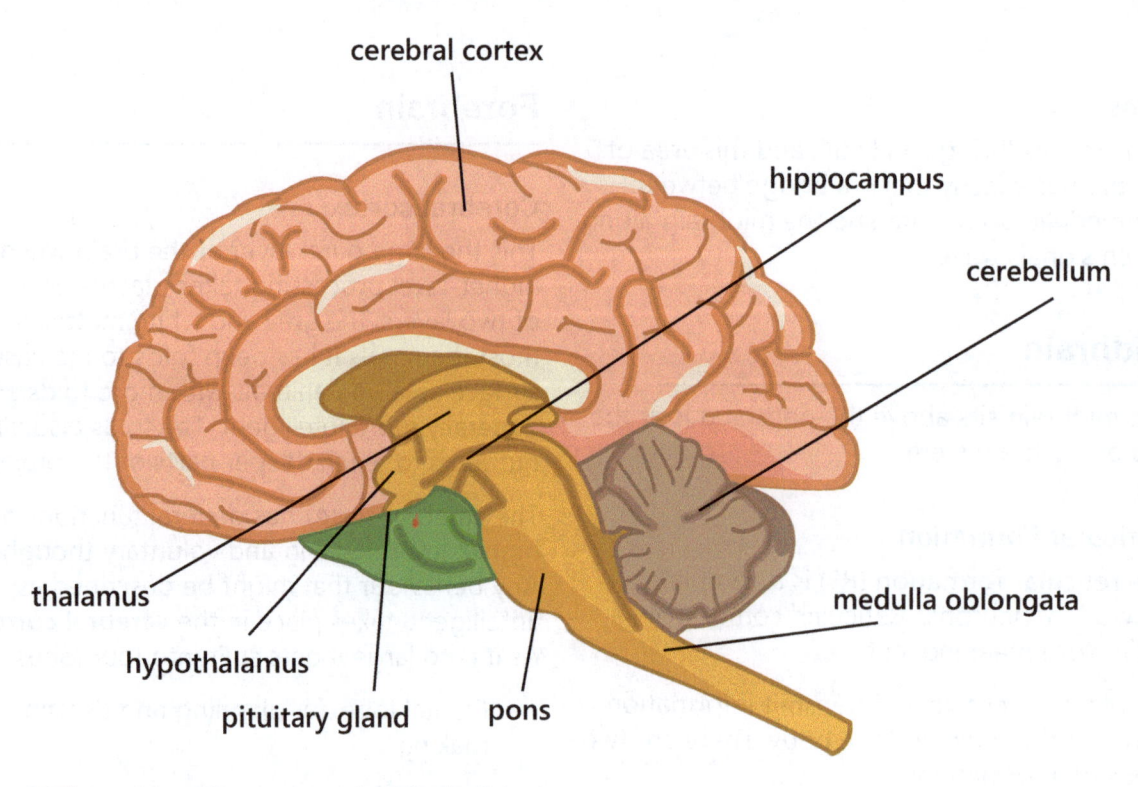

Figure 37 The structures in a cross section of a dog's brain

- The **hypothalamus**: This is a small area of the brain which links the central nervous system to the endocrine system. It is responsible for the homeostasis of a number of bodily functions, such as body temperature, heart rate, and level of hunger. It does this by releasing hormones which stimulate the pituitary gland, which then goes on to produce more hormones which are sent around the body to control different organs. The hypothalamus also controls the body's physical responses via hormones – for instance, fear will cause the hypothalamus to stimulate the endocrine system to produce adrenaline and cortisol.

- The **pituitary gland**: This gland sits beneath the hypothalamus. It produces its own hormones, and stores and releases hormones that the hypothalamus produces. It also stimulates other glands in the body.

- The **amygdala**: This is the emotion centre of the brain, which processes information and attaches an emotional meaning to it. It also helps detect potential threats from the environment and, if required, prepares the animal for the 'fight or flight' response.

- The hippocampus: this is the area where new events are processed and stored as long-term memories. An emotional response to an event partly draws on past experience, which is why the hippocampus plays an important role in the limbic system. The name 'hippocampus' comes from its seahorse-like shape.

- The cingulate gyrus: This has a role in reacting to pain and regulating aggression, as well as linking sights and smells with memories.

- The basal ganglia: This area's functions include working memory and attention, as well as inhibiting undesired movements.

The nervous system

The nervous system controls many of the body's functions. It allows information from the brain to travel to every part of the body, and vice versa. It can be broken down into three separate but related systems:

- the central nervous system (CNS)
- the peripheral nervous system (PNS)
- the autonomic nervous system (ANS)

The nervous system as a whole is made up of countless specialised cells called neurons and glia.

Central Nervous System

The **central nervous system** (CNS) is made up of the brain and the spinal cord. The brain controls and coordinates everything in the body. The spinal cord is attached to the brain. Signals travel to and from the brain and body through the spinal cord.

The brain is protected by a bony skull and cerebral spinal fluid, which acts like a shock absorber. Three layers of connective tissue called **meninges** hold the brain in place within the fluid. The brain is protected from infection by the blood-brain barrier. This stops any toxins or pathogens in the blood from entering the brain. The spinal cord is protected by the vertebrae in the spinal coumn.

Neurons

A **neuron** is another name for a nerve cell. Neurons are specially adapted to transmit and receive electrical and chemical signals. Different neurons have different functions (see next section), however all neurons are made up of the following:

- **nucleus** – Like most other cells, a neuron has a cell nucleus which contains the animal's DNA
- **cell body/soma** – Contains the nucleus and other important structures, and controls the neuron.
- **dendrites** – These receive signals and send them towards the cell body/soma. They have a large surface area with tree-like extensions.
- **axon** – Transmits electrical signals away from the cell body/soma. The larger the axon the faster the message travels.
- **Schwann cell** – These form a myelin sheath which insulates the axon and helps speed up the electrical signal. (Myelin is a fatty protective covering.)
- **nodes of Ranvier** - Gaps in the myelin around the axon that allow nutrients and oxygen into the axon. The gaps also help speed up the electrical signal.
- **axon terminal** – the axon ends in numerous axon terminals

The axon terminals at the end of one neuron form junctions, separated by a tiny gap, with the dendrites of other neurons. This junction is called a **synapse**. Electrical signals that travel to the axon terminals stimulate the release of **neurotransmitters**. Neurotransmitters are stored in membranes called **synaptic vessicles**.

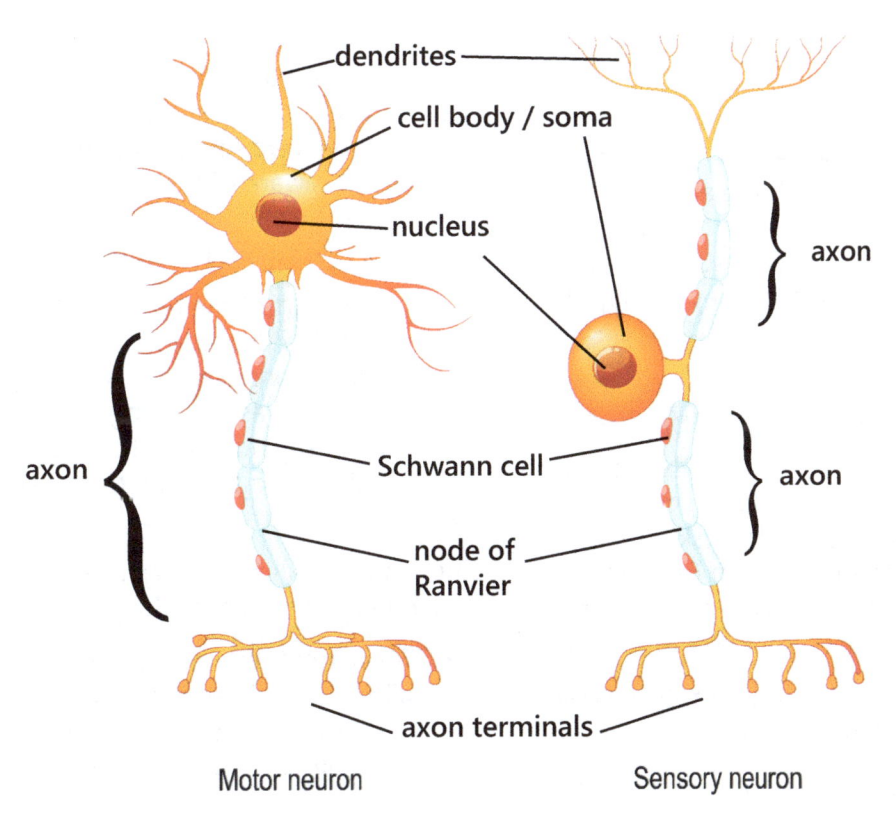

Figure 38 An efferent (motor) and afferent (sensory) neuron

3.2 Neural control mechanisms in animals

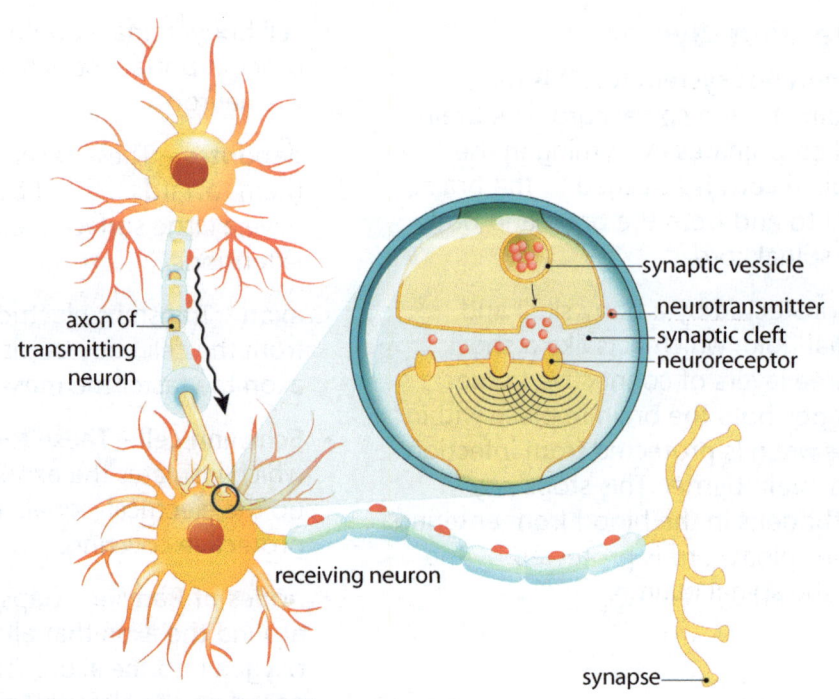

Figure 39 A synapse - note the synaptic vessicles and synaptic cleft

Neurotransmitters are special chemicals that travel across the synapse gap (called the **synaptic cleft**) and bind to **neuroreceptors** on the dendrites of the receiving neuron. This signal travels down the dendrite of the receiving neuron and on to its axon.

Synapses, neurotransmitters and neuroreceptors allow signals to travel between neurons. This process is how the nervous system works.

Glia

The other major component of the nervous system are glial cells which are also just called glia. Unlike neurons they do not carry electrical signals; instead their role is to surround and protect neurons, and ensure they are supplied with sufficient nutrients.

Peripheral Nervous System

The **peripheral nervous system** (PNS) connects the central nervous system to all parts of the body, such as the limbs and the skin. The PNS is made up of a network of glia cells and neurons, which carry signals to and from the CNS.

Unlike the CNS, the PNS is not protected by bone and is therefore more easily damaged.

The PNS makes use of two types of neurons:

- Afferent or sensory neurons carry information from sense organs (e.g. the skin) towards the CNS. (The A in Afferent is because they Arrive at the CNS),

- Efferent or motor neurons carry information away from the CNS to produce a response. Efferent neurons attach to muscles, to move them, or to glands to make a hormone. (The E in Efferent is because they Exit the CNS.)

There are two parts to the peripheral nervous system. The first is called the somatic nervous system. It delivers sensory information which an animal is aware of (using afferent neurons), and allows the animal to voluntarily control its body (using efferent neurons).

The second part of the periperhal nervous system is called the **autonomic nervous system** (ANS) It is responsible for involuntary bodily functions – such as breathing, heart rate and pupil size. Many homeostatic processes are controlled or monitored by the autonomic nervous system.

There are two divisions of the ANS that have separate but complementary functions.

- The **sympathetic nervous system** is also called the 'fight or flight' system. (Think S for Stress). It acts to prepare the body for imminent exertion, and is a response to stressful situations where the animal perceives it is in danger. It does this by increasing the heart and breathing rate, and directing blood towards skeletal muscles and the brain, to run, fight or think its way out of danger. The liver also releases any stored glucose, so that skeletal muscles have access to as much energy as possible. Pupils dilate to get more sense information to the brain. Blood is moved away from smooth muscle, meaning that the excretory, digestive and immune systems are inhibited.

- The **parasympathetic nervous system** is also called the 'rest and digest' system. (Think P for Peace). It acts in the opposite way to the sympathetic nervous system. The parasympathetic nervous system acts to conserve energy and allow digestion to occur. The heart rate and breathing rate is slowed, and pupils and muscles relax. Blood is returned to the smooth muscle, allowing the excretory, digestive and immune systems to work as normal allowing food to be digested.

These two systems are constantly working in opposition, so that animals can respond in the best way to their environment and circumstances.

As an example consider a deer grazing in a forest. It suddenly catches the scent of a very dangerous animal - a wolf. As soon as the brain recognises that the scent represents a threat, the sympathetic nervous system immediately prepares the body for intense physical exertion:

- The pupils dilate.
- Breathing rate increase rapidly, to bring as much oxygen into the blood as possible.
- Heart rate increases rapidly, to ensure that oxygen is distributed as quickly as possible and available for all muscles to use.
- The liver converts stored glycogen into glucose, which increases blood sugar levels, as another source of immediate energy.
- The adrenal glands produce adrenaline (to make the heart beat faster) and cortisol, which triggers the production of yet more

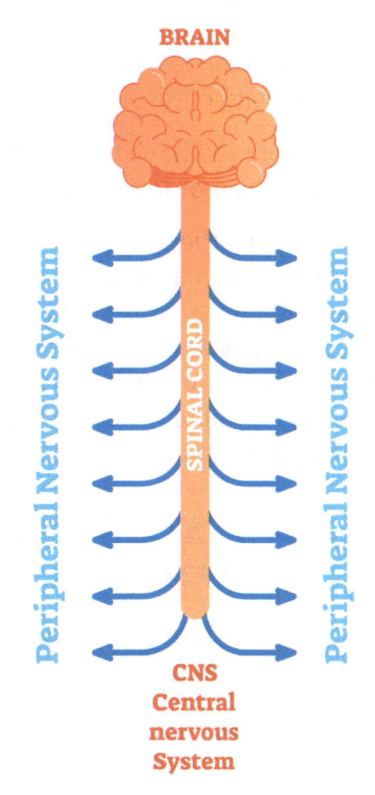

Figure 40 The central nervous system and peripheral nervous system

blood sugar, using another (non-glycogen) process.

- Digestion slows down, and blood is diverted away from the digestive organs. (The next time you do some vigorous exercise, put your hand on your stomach - you will notice it is cold, because blood has been diverted away from it).

This happens very quickly. The deer is on red alert and has diverted all resources available to it's muscles. Wisely, it decides to run. With the amount of energy at its disposal the deer is able to run faster, for longer, than it ever has.

Luckily, the deer escapes. Soon, it no longer has the wolf's scent. With the danger over, the parasympathetic nervous system takes over. It needs to act because the energy available for normal bodily functions is now very low. It reverses all previous actions:

- Pupils return to their normal size.
- Breathing slows down.
- Heart rate slows down.

- The liver stops converting glycogen to glucose and reverses the process by converting any spare glucose into glycogen which it stores.
- The adrenal glands stop producing adrenaline and cortisol.
- Blood is diverted back to all areas of the body; digestion of food resumes.

Without the sympathetic system, the deer would not have been able to run fast or far enough to escape the wolf. Without the parasympathetic system, the deer's metabolic processes would be in constant overdrive - it would quickly use up all energy reseves but would not be able to digest food.

Figure 41 When a deer senses danger, the sympathetic nervous sytem is called into action

Questions

1. Label the structures shown in the diagram. (4)

2. Explain the role that neurons play in passing signals along the nervous system. (3)

3. Name the three structures that make up the hindbrain. (3)

4. It is feeding time for a lion in a zoo. A piece of meat is lowered into the lion enclosure. Outline the main parts of the brain that are involved in the lion sensing, moving towards, eating and digesting the food. (6)

5. Describe the function of motor neurons. (2)

6. What kind of neurons make up the afferent peripheral nervous system? (1)

7. Explain why the autonomic nervous system is split into the sympathetic and parasympathetic nervous systems. (4)

8. Label the structures shown in the diagram. (4)

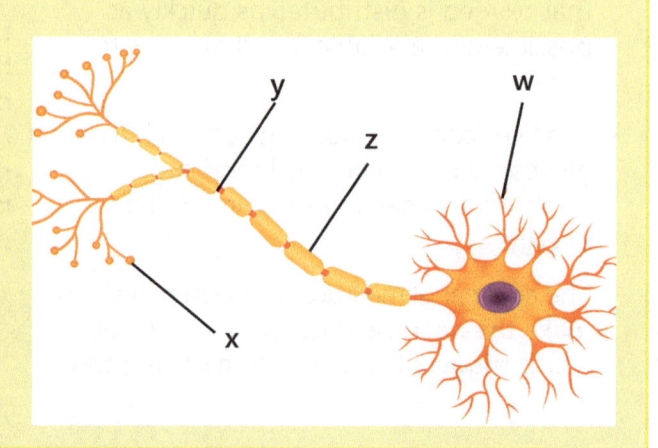

Learning Outcome 3 – Revision checklist

3.1	☺	😐	☹
I can describe the anatomy, and identify functions, of the forebrain, including: the thalamus, hypothalamus, cerebral cortex, limbic system.			
I can describe the anatomy, and identify functions, of the midbrain, including: the reticular formation and the role of neuron receptors.			
I can describe the anatomy, and identify functions, of the hindbrain, including: the medulla, cerebellum and pons.			
I can locate the approximate positions of these components of the brain on a diagram.			

3.2	☺	😐	☹
I can describe the main components of the Central Nervous System.			
I can describe the main components of the Peripheral Nervous System.			
I can identify the different functions of the afferent and efferent systems.			
I can describe the main components of the Autonomic Nervous System.			
I can identify the different functions of the sympathetic and parasympathetic nervous systems.			
I can identify the Central Nervous System and the Peripheral Nervous System on a diagram.			

Learning Outcome 3 - My Revision Goals

Consider each of the points in the revision checklist. For each point, what do you still need to do to make sure you are fully prepared for your exam?

On a separate piece of paper, or using the space below, write down each of the tasks you identify. (If this book was given to you by your tutor then ask before writing in it!)

LO3 Understand the neural control mechanisms in animals

LO4 Understand how animals' senses have adapted to their environment

4.1 How animals' senses are adapted to their environment

In this topic you will learn about:

- The eye.
- The ear.
- The nose.
- The mouth.
- The sense of touch.

Whilst animals share many characteristics – for instance, all mammals have a number of important features in common – each individual species is the result of a series of evolutionary adaptations. These are adaptations to the environment in which they live. These adaptations take place over very long periods of time.

The sense organs allow an animal to experience and interact with the world. They are the means by which an animal perceives danger, food and potential mates. Adaptations to sense organs can give a species a small advantage in their battle to survive, find food and reproduce. Indeed, these small advantages are the reason for such adaptations.

There are five major senses:

- vision – detects light using photoreceptors
- hearing – detects air pressure changes using mechanoreceptors
- smell – detects chemicals using chemoreceptors
- taste – detects chemicals using chemoreceptors
- touch – detects pressure using mechanoreceptors

The sense organ dedicated to each sense is discussed below.

Link
There are also other specialised senses and sense organs that will be discussed in section 4.2.

Eye

The eye is adapted to detect light. When it does it generates electrical signals which are sent to the brain for processing. The main features of the eye (Figure 42) are:

Outer layer

- **cornea** – A thin, transparent disc at the front of the eye that bends (refracts) light through the pupil and lens so that it focuses on the retina. Whilst the lens fine-tunes this bending, most of the bending takes place in the cornea. The cornea also protects the front of the eye from damage.

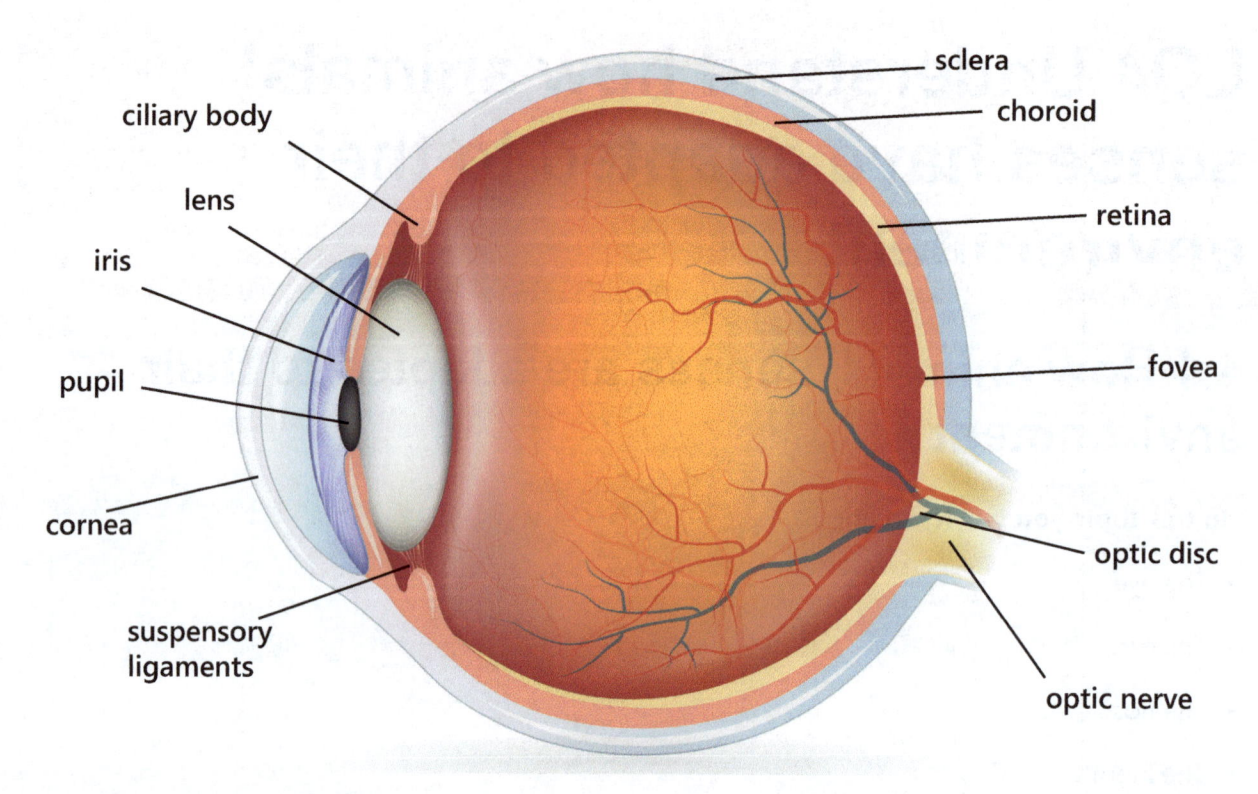

Figure 42 The structure of the eye

- **sclera** – The tough outer layer of the eye, made of connective tissue and visible as the 'whites of the eye'. It protects the eye and keeps it in shape.

Middle layer

- **iris** – This is a ring of muscle (sphincter) which makes up the coloured part of the eye. It can dilate (widen) and constrict (narrow) to let more or less light pass through the pupil, in order to provide optimal vision. This widening and narrowing is involuntary and depends on how light or dark it is.

- **pupil** – This is the opening in the iris which lets light through. It is completely black because the light that enters is absorbed within the inner eye.

- **ciliary body** – this ring of tissue at the front of the eye is connected to the lens and contains a muscle which can change the shape of the lens. It is also responsible for producing aqueous humor, a liquid which provides nutrients to the lens and the cornea.

- **lens** – The lens is a concave shape, made of transparent and flexible tissue. The lens changes shape to bend light beams so they focus on the retina. Light from far away objects need less bending than light from nearby objects (Figure 43). The image focused on the retina is upside down but the brain turns the image over when being processed.

- **suspensory ligaments** – These support the lens.

- **choroid** – A layer of tissue between the retina and sclera that contains lots of blood vessels that provide nutrients and remove waste from the eye.

Inner layer

- **retina** – A thin layer of cells at the back of the eye, made up of photoreceptors (light-sensitive cells) called rods and cones. They react to light hitting them by sending an electrical signal through the optic nerve to the brain.

- **rod cells** – Rod cells are photoreceptors that are very sensitive to light, so they work well in low light levels. They are not sensitive to different colours. The image they form is blurry - it is not fully focused.

- **cone cells** - Cone cells are photoreceptors

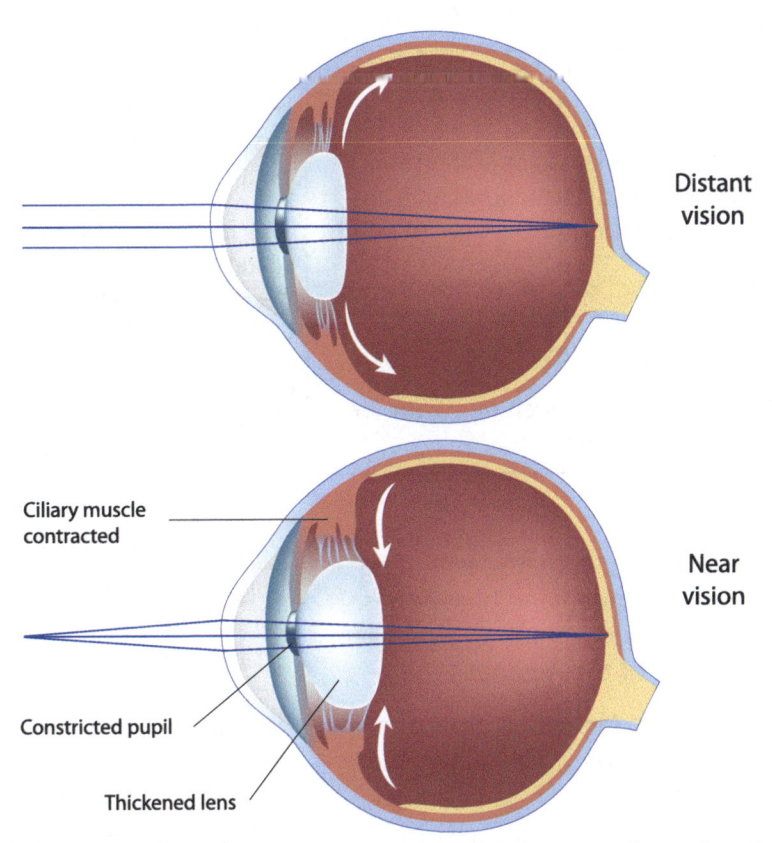

Figure 43 The lens changes shape to ensure light from both near and distant objects are focused on the retina. For distant vision the lens becomes thinner. For near vision the lens becomes thicker.

that are responsible for colour vision and fine focus, so they form clear images. They are not as sensitive to light as rod cells so they need more light to function.

- **fovea centralis** – This is a small pit in the retina that only contains densely-packed cone cells; it is responsible for the sharpest, most detailed vision.
- **tapetum lucidum** - A thin layer of reflective cells behind the retina that reflects light back into the eye. It helps animals to see in low light levels. It makes animals' eyes glow or shine in the dark. Many mammals and other nocturnal animals have a tapetum lucidum. However, most primates and pigs do not.
- **optic disc** – This is the point where the optic nerve begins, and leaves the eye; there are no light-sensitive cells (rods or cones) at the site of the optic disc, and therefore this part of the eye does not detect light – it is the location of a small 'blind spot'.
- **optic nerve** – The optic nerve is made of neurons which transmit electrical signals from the retina to the brain.
- **medial** and **lateral rectus muscles** – not shown on the diagram, these muscles sit on the left and right of each eye; the medial rectus moves the eye to look inwards (towards the nose), and the lateral rectus moves the eye to look outward (Note: other muscles move the eye up and down).

General adaptations

Different species have developed adaptations to their sight, based on their environment and lifestyle. Human eyesight is also the result of adaptations, meaning our eyesight is different to other animals.

Colour vision depends on the characteristics and number of cone cells:

- Two types of cone cells: Most mammals have blue and green cones but cannot discriminate between reds and greens. Most mammals' colour vision would

Predator species	Prey species
Often have eyes on the front of the head and see almost the same image. This allows for 3D binocular vision – the brain can combine the slightly different images from each eye to build up an accurate sense of depth. This allows predators to perceive where objects, including prey, are in three dimensions. This allows them to pounce on prey. However the field of view is much narrower for animals with binocular vision.	Often have eyes on side of head to give a wide field of view so that predators cannot sneak up unseen. Some species have a small blindspot in front or behind them. This is known as panoramic vision or monocular vision. However, panoramic vision means that the image in each eye is vastly different and the brain cannot combine these into one 3D image. Panoramic vision normally means a lack of 3D binocular vision. This means that depth perception is generally not as good as predators.

appear washed out and drab to us.

- Three types of cones: Primates, including humans, have red, green and blue cones. (Colour blindness is the result of a problem with one or more of these cones.)
- Four types of cone: Almost all birds and fish have a fourth cone which allows them to see ultraviolet.
- Five types of cone: Some animals, including pigeons and some reptiles, have a fifth cone which gives their vision a vibrancy that we cannot imagine!

The cones of birds, and some reptiles and fish, are even more sensitive than mammal cones, and are referred to as 'double cones'. They also have special 'oil droplets' which form part of the cone structure, which act as light filters. All of these structures together account for the high level of vision that is found in these animals.

Bird and reptiles also have another structure, called a nictitating membrane, which is like a third eyelid used for blinking and lubrication.

Nocturnal animals have a few adaptations to help them see in the dark:

- They have more rods than cones, as these work well in low light conditions
- They have large pupils to let in as much light as possible
- They also have slit pupils to prevent light from damaging the eye. Slit pupils work with a multifocal lens which is arranged in rings, each of which bends different frequencies of light. The multifocal lens helps keep clear focus at all light frequencies. The slit pupil allows light to hit each ring of the lens whilst still restricting the overall amount of light.
- They have a tapetum lucidum as discussed earlier.

A bit more about light

Light is a vibration of electric and magnetic fields. When these fields vibrate at a particular rate (frequency) animals' eyes can sense it. This is really what we mean by 'light'. Cone cells can detect specific vibration frequencies, which are perceived as different colours. But the electric and magnetic fields vibrate at many other frequencies too. Ultraviolet, infrared, radio waves, microwaves and even X-rays are all vibrations of the electric and magnetic fields. The difference is that our eyes cannot see them, and hence they are 'invisible' to us. Some animals, however, can detect vibration frequencies that are invisible to us – for instance, many birds can see ultraviolet light.

Some other examples of adaptations are shown in the table on the opposite page.

Figure 44 The nictitating membrane in a bald eagle

Owl	Goat
- A nocturnal animal such as an owl has excellent night vision – due to the density of rods - However, owls have more rods than cones, so they have less good colour vision – but they don't need to see lots of different colours in the dark - Very effective 3D (binocular) vision as their eyes are on the front of their face – necessary to accurately locate and swoop on prey	- Unusual rectangular and horizontal pupils provide very wide panoramic vision – a predator cannot sneak up unseen
Birds of prey	**Cat**
- Among the best eyesight of any animal - Have a very high density of rods and cells - Eyes are large relative to their body, to let lots of light in - Often have two fovea – which means they can focus on two different things at once - Some birds can see in ultraviolet	- Cats are crepuscular (i.e. most active at dawn and dusk) and their eyes have adapted to this. - Their pupils are able to open very wide, to let in lots of light, but can also narrow to a slit in daylight conditions - They can do this because their ciliary body is made up of two muscles instead of one - They have relatively poor colour vision – because they rely on detecting movement more than colours - Cats have poor vision very close to their face but use their whiskers and sense of touch to help detect objects, including prey
Snakes	**Arctic Reindeer**
- Pythons, vipers and boas are able to detect infrared light which helps them hunt at night. Infrared allows the snake to sense the heat of their prey	- Can see in ultraviolet – important as predators are often camouflaged white against the snow

4.1 How animals' senses are adapted to their environment

Ear

Sounds are vibrations that travel through air, water or solids. It is these vibrations that the ear has evolved to detect. An animal hears by detecting these vibrations and turning them into electrical signals which are sent to the brain. The brain then interprets these signals as hearing.

Sound vibrations in water are different to those in air, which is why things sound different in water.

There are three parts to a mammal's ear: the outer ear, the middle ear and the inner ear. The following describes the main components of each section.

Outer ear

- **Pinna** – The visible portion of the ear whose shape helps to direct sound into the ear and amplify the important pitches. Many animals have muscles to move the pinna in the direction of the sound. The pinna can also have a role in communication - animals such as horses and rabbits use the position of the ear to express their mood.

- **Auditory meatus** (ear canal) – This is a passageway that connects the pinna to the middle ear. It channels sound through to the tympanic membrane. It is covered with hairs and oils to prevent foreign bodies entering the body through the ear.

Middle ear

- **Tympanic membrane** (ear drum) – This is a very thin membrane stretched tight like the surface of a drum. It is the start of the middle ear. Sounds cause the surface of the tympanic membrane to vibrate. It passes these vibrations on to the ossicles (ear bones) and through to the inner ear.

- Ossicles – The three middle ear bones listed below. .

    * **Malleus** (hammer) – The first of three tiny bones in the ear, the malleus is

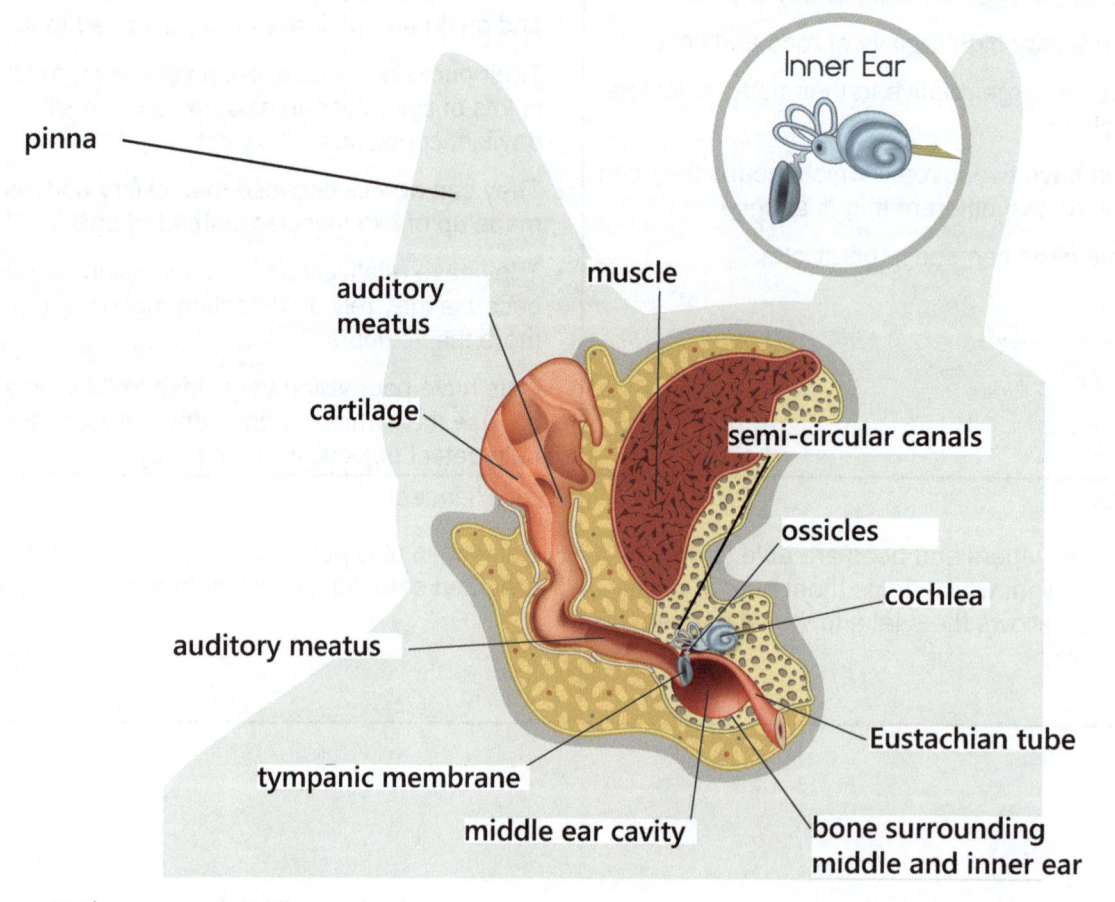

Figure 45 The outer and middle ear of a dog

hammer-shaped and attached to the tympanic membrane.

* **Incus** (anvil) – The second of three bones, the incus is an anvil-shaped bone that receives vibrations from the malleus and passes them to the stapes.

* **Stapes** (stirrup) – The third of three bones, the stapes is a stirrup-shaped bone that passes vibrations from the incus to the oval window in the inner ear.

> **Link**
> In section 1.5 we discussed the mammalian skeleton. The three bones in the ear are unique to mammals - the incus and malleus are evolutions of bones located in reptile jaws.

Vibrations from the tympanic membrane are transmitted by the ossicles to the oval window. The middle ear acts like an amplifier – as the vibrations pass through all three bones, they are amplified by 10. The amplifying process increases hearing sensitivity

The middle ear is also connected to the throat via the Eustachian tube. It keeps the pressure in the middle ear the same as the external environment. It does this by opening with the help of muscles used for swallowing and yawning. This sensation can be felt when your ears 'pop'.

Inner ear

- **Oval window** – A thin membrane that connects the stapes to a fluid-filled cavity within the cochlea. It connects the middle ear to the inner ear. The vibrations from the stapes causes the oval window to vibrate.

- **Round window** – This is another membrane-covered opening of the cochlea. Its function is to allow the fluid within the cochlea space to move. So, for instance, when the oval window membrane is pushed into the cochlea by the stapes, the round window membrane is pushed outward by the movement of the fluid. If the round window were not present then the oval window would not be able to move the fluid in the cochlea, and thus hearing would be lost.

- **Cochlea** – The cochlea works with the organ of Corti to convert vibrations into electrical signals. In mammals the cochlea is a spiral structure that looks like a snail shell and is filled with a fluid which vibrates when the oval window vibrates. Wrapped up in the middle of the cochlea shell is a membrane. The membrane starts off narrow by the oval window but gets wider and wider towards the centre (apex) of the spiral. Slow vibrations in the fluid (low sounds) only cause the wide part of the membrane to move. Fast vibrations in the fluid (high sounds) only cause the narrow part of the membrane to move.

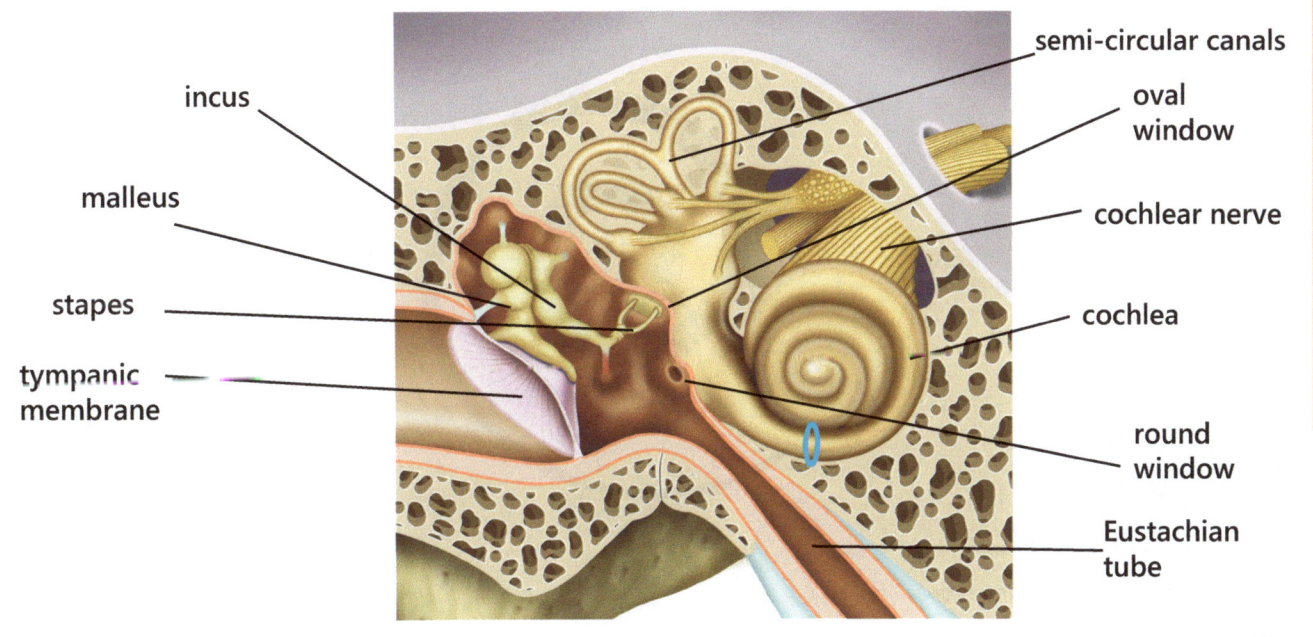

Figure 46 The structure of the middle and inner ear. The blue semi-circle is the cross-section which is expanded in Figure 47.

LO4 Understand how animals' senses have adapted to their environment

a cross-section of a cochlea spiral

organ of Corti

membrane

hair-like projections

cochlear nerve

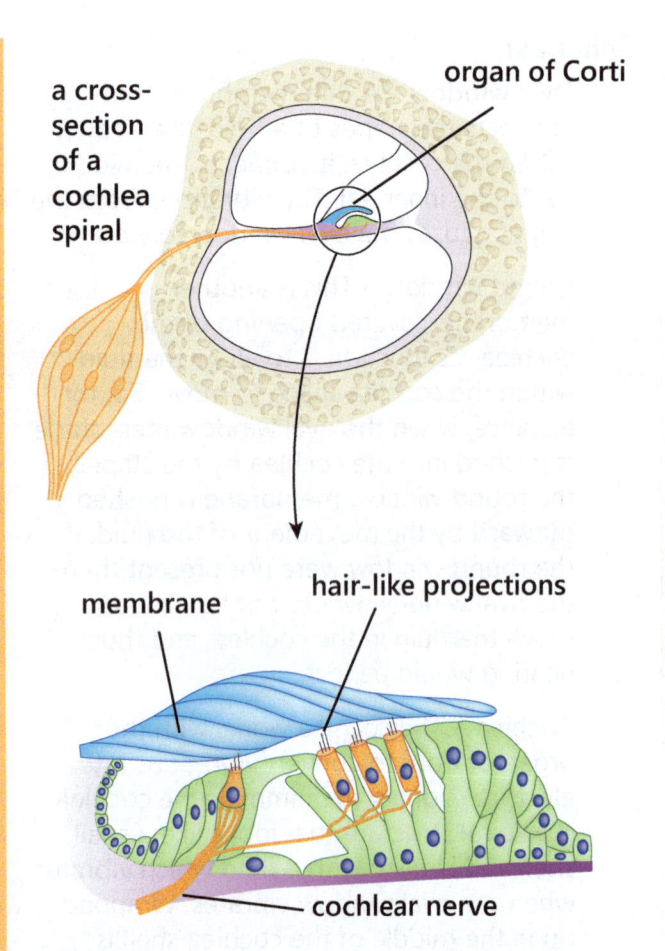

Figure 47 The cochlea and the organ of Corti. The top image shows a cross-section through the cochlea spiral

- **Organ of Corti** – This is attached to the membrane in the cochlea and consists of special cells with hair-like projections, all along the membrane. When stimulated by movement of the membrane, these hairs send electrical signals to the cochlear nerve. Slow vibrations (low sounds) are detected in the wide part of the membrane, in the middle (apex) of the spiral. Fast vibrations (high sounds) are detected in the narrow part of the membrane, in the base of the spiral. (See Figure 47a). Electrical signals from different cells correspond to different sound frequencies. To summarise:

  * Low-pitch sounds are detected in the centre (apex) of the spiral, far from the oval window.

  * High-pitch sounds are detected at the base of the spiral, near the oval window.

- **Cochlear nerve** – This transports the electrical signals from the organ of Corti to the brain for interpretation.

- **Semi-circular canals** – This part of the ear controls balance and has nothing to do with hearing. The canals act like a spirit level for the animal's body. Each of the three interconnected canals has a different orientation (up/down, left/right, forward/backwards and each is filled with fluid. The movement of fluid in each stimulates hair-like protrusions, just like in the organ of Corti, which generates electrical signals which are sent to the brain. The differences between the signals from the three canals are interpreted by the brain and used to coordinate position and balance.

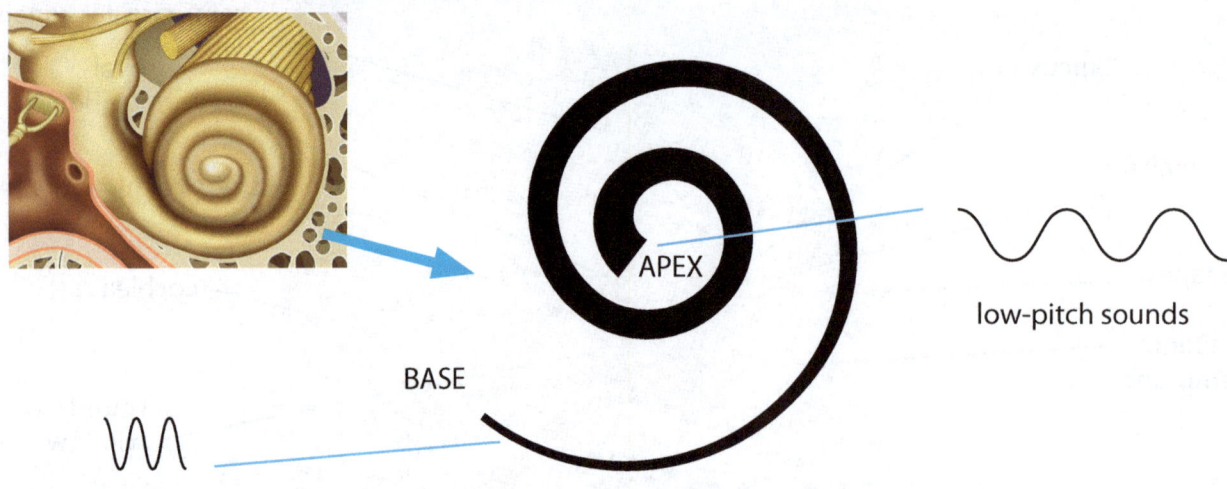

APEX

BASE

low-pitch sounds

high-pitch sounds

Figure 47a The cochlea membrane

Predator species	Prey species
Ears are often forward-facing – slight differences between the sounds in each ear allows them to pinpoint the exact location of the sound.	Most prey species have movable ears, so that the animal can point their ears in a range of directions to ensure nothing sneaks up on them.

General adaptations

The internal mechanism of the ear, as described above, is the same for all mammals. The malleus and incus evolved from bones in the reptile jaw and are specific to mammals. Non-mammals' ears work in slightly different ways.

The most obvious feature of the mammalian ear is the pinna. Birds, fish and reptiles do not have a pinna but they do normally have ears.

Even within mammals the pinna can look quite different across species. In each case these evolutionary differences are due to habitat and lifestyle.

The various shapes and sizes of pinna help to ensure that the most important sounds for that animal are passed down into the middle ear. However, the pinna also has a role in regulation of body temperature (see section 2.3 for more on thermoregulation). For instance, an elephant's large pinna do not aid its hearing but instead help it to keep cool. The elephant's pinna have a large network of blood vessels close to the skin surface that helps them get rid of any excess heat and cool them down.

Nose

The sense of smell allows animals to detect substances in the air. It does this by reacting to the chemicals that make up these substances. These are either inhaled in through the nostrils or via the throat when eating food. (The sense of smell also helps inform how things taste.) The sense of smell is known more formally as the olfactory system.

The olfactory system is made up of the following structures:

- **Nasal chambers** – this refers to the spaces leading from the nostrils all the way to the top of the throat. Each nostril leads to its own chamber, separated from the other by the septum, until the two chambers merge at the top of the throat. The turbinates and

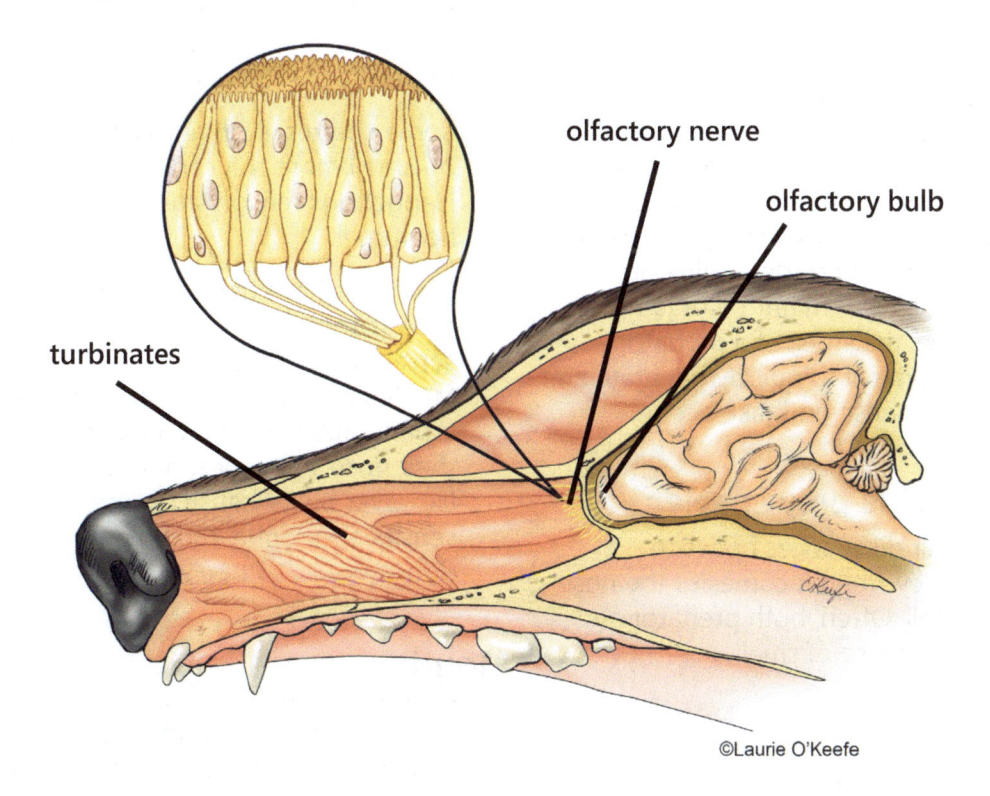

Figure 48 The nose of a dog © Laurie O'Keefe, reproduced with kind permission

Predator species	Prey species
Predators that hunt mainly using smell are finely attuned to the smells associated with their prey. They understand that they need to hunt downwind of their prey so as not to alert them.	Prey species might have an excellent sense of smell, fully attuned to the odour of predatory species – but they might also use some other methods to detect predators and avoid capture.

olfactory nerve are contained within the nasal chambers. The mucus in the nasal chambers aids with olfactory reception by making air moist. Dogs' breathe in and out using separate nostrils.

- **Turbinate bones** – A network of bones, tissue and blood vessels, and also filled with olfactory receptors with a large surface area. Chemicals in substances attach to the receptor cells and trigger an electrical message to pass up the olfactory nerve. Shaped like shells or scrolls, the air passes through the turbinate bones during breathing. They warm and moisten the air, and regulate the moisture content of the air that is exhaled. They can do this because blood vessels in the turbinates are very close to the surface.
- **Olfactory nerve** – Travelling from the back of the nose up to the olfactory bulb, the nerve passes electrical signals to the olfactory bulb.
- **Olfactory bulb** – The electrical signals reach the olfactory bulb, where smells are processed. The olfactory bulb is able to organise and categorise different smells but sends on information to the cerebral cortex in the brain for further processing. This is so the brain can determine higher-level responses that might be required – for instance, the smell of a predator requires the animal to quickly generate an emotional response and take action.

Predatory versus prey species

There are no clear differences between predatory and prey when it comes to the nose and sense of smell. Often, both predator and prey have a good sense of smell.

General adaptations

Smell evolved as a way for animals to sense the presence of substances in the environment. It was beneficial to be able to detect the presence of food without being able to see it. It was also beneficial to sense when something was potentially toxic or dangerous. Some of the smells we find most repulsive are those associated with food that has gone off and which is, therefore, dangerous to eat.

- Species evolved their sense of smell according to the demands of their ancestors' environment and lifestyle.
- The ability to smell allows animals to communicate with each other by producing special chemicals called pheromones. These are used to signal information about territories, social hierarchies, mating behaviour or the presence of food. For example the male silk moth uses pheromones to find a female and he can detect her up to 7 miles away. The male silk moth has large antennae packed with pheromone receptors. Mammals often use their flehmen response to detect pheromones - see section 4.2.
- Some species, such as hedgehogs, rely on smell more than any other sense. Hedgehogs rely on smell because they are nocturnal.
- Mammals have generally evolved the best sense of smell. Many have a much better sense of smell than humans. The most familiar examples are dogs, who have around 300 million olfactory receptors in their nose - 60 times more than humans. However bears' sense of smell is even more sensitive - around 100 times better than a dog.

Mouth

The mouth is an opening through which most animals take in food. Animals evolved a sense of taste as a way to identify nutritious food and

Figure 49 The left-hand image of a tiger shows the hard palate (towards the very front of the mouth) and soft palate. The right-hand image of a dog shows taste buds on the tongue, visible as small indentations on the surface.

avoid potentially dangerous food – for instance rancid meat or poisonous plants. The sense of taste is often located in the mouth – although this is not the case for all species. The sense of taste is also known as the gustatory system.

Taste buds contain a collection of taste receptor cells, called gustatory cells. They contain hair-like protrusions which are activated by particular chemical compounds in food and send electrical signals through the nervous system to the brain. Taste buds are often located in the mouth, particularly on the tongue and soft palate.

There are different taste receptor cells that respond to particular tastes. Humans have receptors than can sense sweet, savoury (umami), bitter, salty, and sour. However not all animals can sense all of these – for instance, a cat cannot taste sweet things because it does not have sweet taste receptors.

The **palate** is the roof of the mouth which separates the nasal chambers from the mouth. It is made up of a hard palate and a soft palate:

- **Hard palate** – This is the hard part of the roof of the mouth, towards the front. It is ridged and made of bones that are part of the skull. In mammals the vomeronasal organ (see section 4.2) is located just above the hard palate.

- **Soft palate** – This is the roof of the mouth towards the back. It is made of smooth muscle which moves to allow large chunks of food down the pharynx and into the oesophagus. It has a mucous lining which secretes saliva, which is the first step in the digestion process. When swallowing, the soft palate closes the passageway at the back of the throat that is connected to the nasal chambers. This stops food from entering the nasal cavity.

General adaptations

- The hard palate is far more developed in mammals (and crocodiles). It allows a mammal to chew, or suckle milk, and breathe at the same time, which is why it is

Predator species	Prey species
Predator species are less likely to have a full range of taste receptors, particularly sweet receptors, because their diet is less varied than a herbivore or omnivore.	Prey species who are herbivores tend to have a larger number of taste buds than carnivores. It is thought this is because they have a more varied diet, and it also allows them to detect and avoid poisonous plants in their diet.
	Omnivores often have more taste buds than carnivores but fewer than herbivores.

4.1 How animals' senses are adapted to their environment

almost unique to mammals. The evolution of a hard palate in crocodiles allows them to stay submerged in water but still able to breathe.

- The gustatory system and taste buds evolved as a way to determine which food was unsafe and which was nutritious. It also allows animals to determine dangerous levels of something – for instance, salty food is not dangerous in very small quantities but very salty food is generally not good for health. Therefore, taste has evolved to dislike overly-salty food.

- Taste is intricately linked to diet. Cats cannot taste sweet things. This is because cats are obligate carnivores, and have no dietary need for simple carbohydrates (sugars). Over time cats evolved, and lost the taste receptors to sense sweet food.

- Cats are carnivores and have around 500 taste buds. Dogs are omnivores and have around 1700 taste buds. Humans are also omnivores and have around 9,000. The cow is a herbivore and has 25,000 taste buds, because they need to know which plants are safe to eat.

- Some fish have an incredible number of taste buds. The catfish has at least 100,000 taste buds. A large number of these taste buds are not on the tongue but instead on the surface of their skin.

It is widely accepted that birds and reptiles have a poor sense of taste in comparison to mammals. However, the five basic tastes, (savoury, sweet, bitter, salty and sour), are very human-centric. Animals have mainly been tested for their responses to these five tastes. It may be that there are other tastes humans cannot detect but which certain species can, because they have taste receptors that humans do not have. Research into animal taste is ongoing.

Touch

The sense of touch is perhaps the most overlooked of all the senses. However, it played a critical role in the evolution of mammals. The social bonds between a newborn and its mother are created and reinforced by the feelings induced by touch.

Touch is also known as the tactile or somatosensory system. It is different to the other senses because it does not sense things at a distance – instead it senses things in direct contact with the body.

There are **skin receptors** within the dermis, near the surface of the skin, which turn information about touching into electrical signals. These signals are sent via the nervous system to the brain. There are three main types of skin receptors (also known as cutaneous receptors):

- Mechanoreceptors detect normal, gentle touches or pressure.

- Thermoreceptors detect temperature, and changes in temperature

- Nociceptors are pain receptors which respond to sensations that are likely to cause damage to the body, sending signals that the brain interprets as pain.

Figure 50 shows the different receptors in the skin. Some can only perform one function whereas others can perform more than one. The root hair plexus, seen in the diagram, is stimulated when individual hairs move. Given that many animals are covered in hair or fur, this provides animals with further tactile information.

There are more skin receptors in some parts of the body than others. This means that some parts of the body are more sensitive than others.

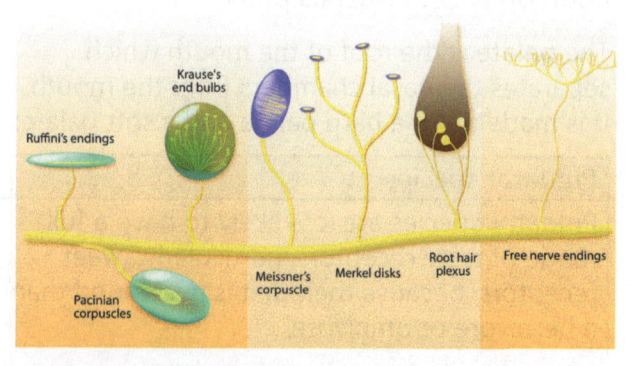

Figure 50 It is thought that different types of skin (cutaneous) receptors in the skin perform different functions. Some of these structures perform more than one function, most notably free nerve endings.

Skin

Large numbers of touch receptors are in the skin. The different layers of the skin are as follows:

- **Hair follicles** – Responsible for hair growth. Hairs insulate most mammals and keep them warm. When animals get cold the tiny muscles at the base of the follicles contract and cause the hairs to stand on end, trapping air next to the animal, which keeps them warm.
- **Sebaceous gland** – These are glands at the base of hair follicles which secrete an oil called sebum to lubricate the skin and hair. For some species these secretions make it water repellent.
- **Epidermis** – The top layer of skin made of epithelial cells. It provides a waterproof layer and also prevents pathogens from entering the animal's body.
- **Dermis** – This layer contains blood vessels (including those responsible for vasodilation and vasoconstriction), glands, hair follicles, nerves and touch receptors. It also contains proteins called collagen and elastin, which provide the animal's skin with structure, strength and flexibility.
- **Hypodermis** (subcutaneous layer) – Made from fat cells (adipose tissue) which insulates the animal from the cold.

Predator versus prey species

Whilst there are no clear general differences between predator and prey species, some species use their sense of touch to either detect potential prey or to detect potential predators. Many insects and arachnids have sensilla, which often look like hairs, to detect movement:

- feline species use their whiskers to help them hunt but some also use them to detect whether prey has stopped moving before releasing their grip
- tarantulas use their sensilla to help navigate, and sense vibrations in the ground as a way to detect prey
- scorpions use sensilla on their pincers for navigation, which can also detect tiny currents of air in order to sense the movement of prey

General adaptations

The main differences across species are the number of receptors in different parts of the body. These differences have evolved according to lifestyle and diet. For example:

- Hairless skin (also known as *glabrous* skin) is generally more sensitive to touch than hairy skin – hence the most sensitive parts of animals' bodies are not covered in hair.
- Primates have very dextrous hands and feet, which they use to explore their surroundings and handle food. Their hands and feet have large numbers of receptors, which provides very high levels of sensitivity to textures and pressures. Conversely, the paws of animals that walk on all fours, such as dogs and cats, are not as sensitive as primates' hands and feet.
- The sense of touch around the nose and mouth is very sensitive for animals like pigs, that use the nose and mouth to explore their environment.

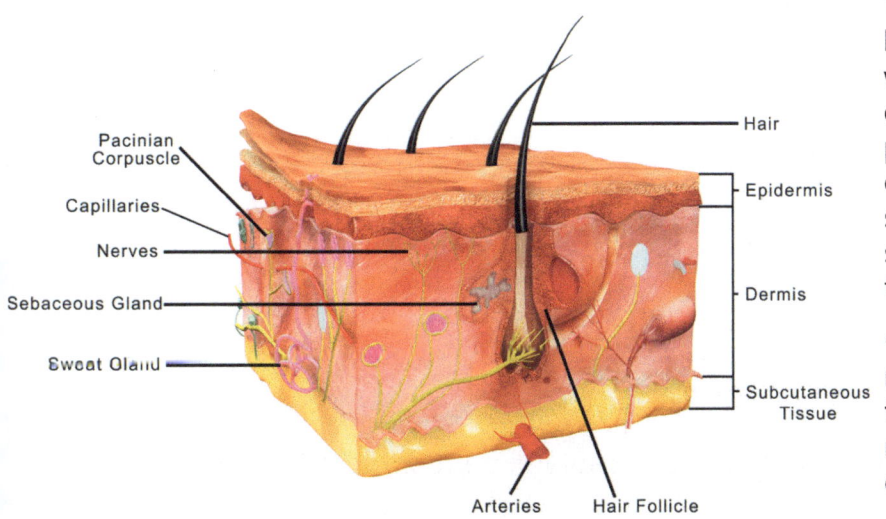

Figure 50a The structure of the skin.

4.1 How animals' senses are adapted to their environment

Questions

1 Label the structures shown in the diagram. (5)

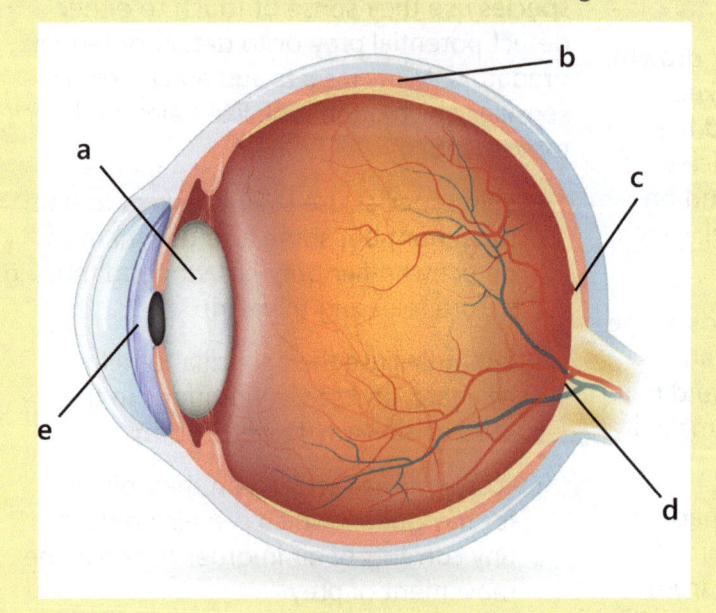

a _____
b _____
c _____
d _____
e _____

2 Describe three adaptations to the eye that might aid a nocturnal predator. (3)

3 What is the role of the choroid? (1)

4 Outline the processes that take place in the eye when a golden eagle searches for and locates prey. (4)

5 Label the structures shown in the diagram. (4)

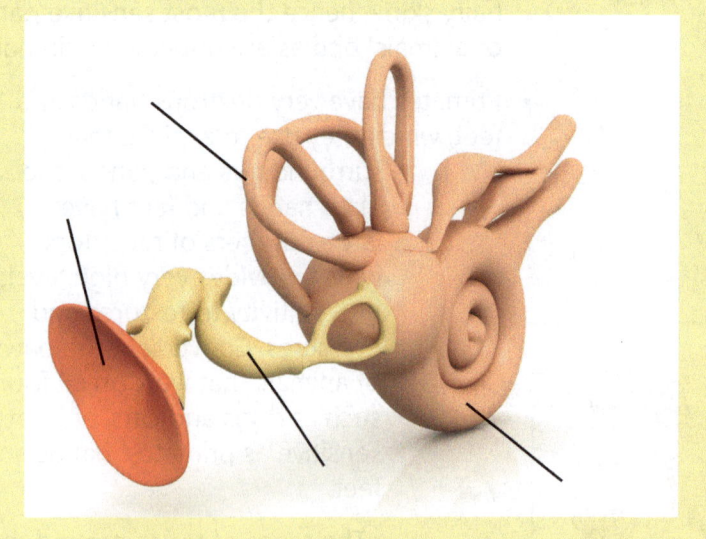

6 What is the role of the organ of Corti? (2)

7 Describe the evolutionary adaptations of a rabbit's ear compared to the ear of a lizard. (6)

Questions (continued...)

8 Label the structures shown in the diagram. (3)

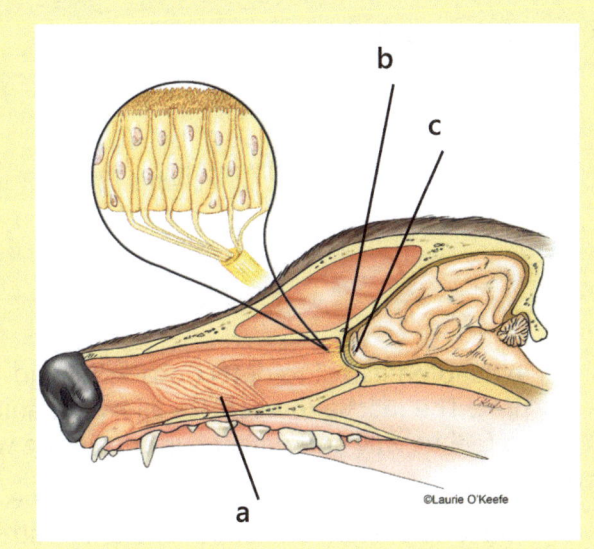

a _____

b _____

c _____

9 What is the role of the olfactory bulb? (1)

10 Explain why dogs have developed a better sense of smell than humans. (3)

11 Outline the main processes involved in the ear when a deer hears an unexpected sound. (5)

12 Describe the role of the palate in a mammal's mouth. (2)

13 Explain why there are differences between the sense of taste in a cat and a catfish. (5)

14 What is the difference between the sense of taste and the sense of smell? (2)

15 Describe two types of skin receptor. (2)

16 What similarities and differences might there be between the sense of touch in a great ape and the sense of touch in a horse? (4)

4.2 Specialised senses

In this topic you will learn about:

- A range of specialised senses and appreciate why they have evolved.
- Specialised tactile organs.
- Specialised taste and smell organs.
- Electroreception.
- Echolocation.

The senses discussed so far are those that animals share with humans. However, some species have evolved different specialised organs for touch, taste and smell, or developed senses that have no equivalent in humans.

Tactile organs

Platypus bill

The duck-billed platypus is a unique mammal for many different reasons but one of them is the bill itself, which is really a sense organ.

The bill is soft and flexible, not hard and rigid like a duck's, and has tens of thousands of mechanoreceptors (as well as electroreceptors – see later). The animal feeds mainly at night on aquatic invertebrates, foraging for them on the bottom of rivers and streams. It closes its eyes, ears and nose whilst hunting, relying instead on its bill to detect obstacles and find prey. The mechanoreceptors allow the animal to feel its way and detect currents in the water.

The duck-billed platypus is a **monotreme**. It is native to eastern and south eastern Australia, including Tasmania. It lives by streams and rivers and is semi-aquatic – it is far better at swimming than walking on land.

> **Link**
> For more on monotremes see section 1.3.

Lateral line

The **lateral line** is an organ that allows fish to detect the movement of water. This helps them to detect prey, avoid predators and swim in a school with other fish. It consists of a series of openings, running down the length of the skin, all connected to the lateral line canal (Figures 52 & 53). The sense organs themselves

Figure 51 A duck-billed platypus

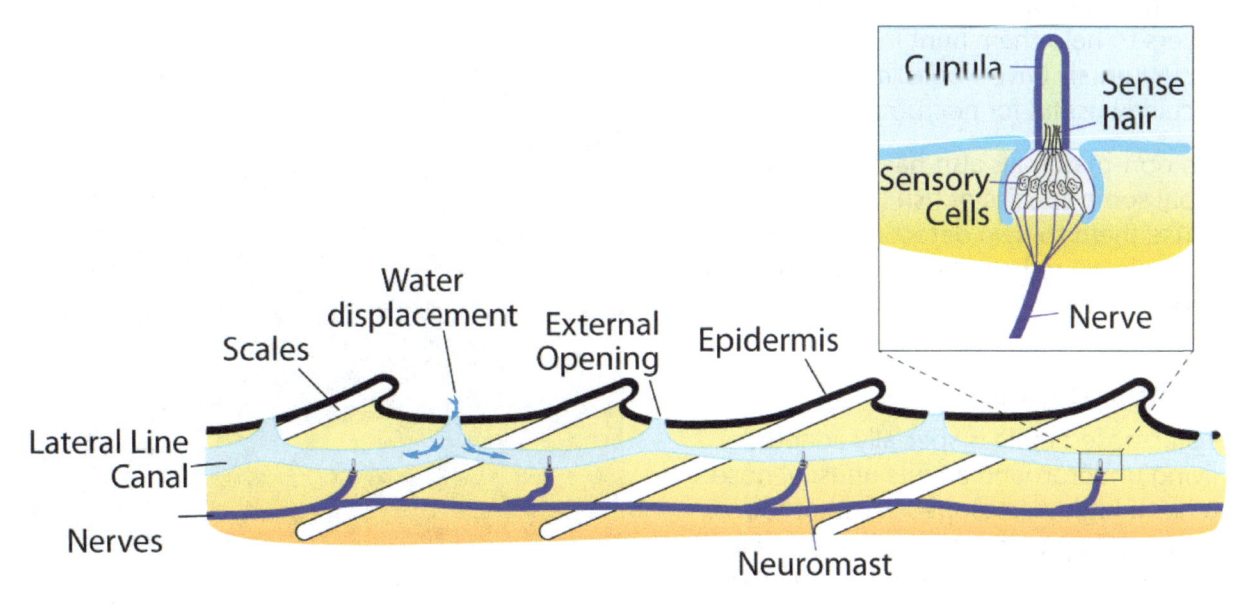

Figure 52 The lateral line. © Thomas Haslwanter.

are called neuromasts, which are hair-like projections which extend into the canal and move with the movements of water in the canal. Movements of these 'hairs' generate electrical signals in the nerve cells to which they are attached. These electrical signals then travel through the nervous system and to the brain.

Did you know?

The system of detecting movement using hair-like projections within a liquid is the same as that used in the organ of Corti in the inner ear.

Vibrissae

Vibrissae are special hairs whose only function is to provide sensory tactile information. They tend to be longer and thicker than normal hair. (Vibrissae is the plural of vibrissa).

The most well-known vibrissae are whiskers. The root hair plexus associated with whiskers provides the animals with further sensory clues about their surroundings. Cats have poor close-up vision (around 20cm from their eyes) so they use their whiskers when hunting to detect prey up close and help them pounce. They also use their whiskers to judge whether

Figure 53 A lateral line in fish can be clearly seen on the surface of the skin

Figure 54 The vibrissae of a catfish

they can fit through small gaps. Seals use their whiskers to help them hunt in murky waters. Most mammals have vibrissae and they are particularly useful for nocturnal animals.

Some non-mammals also have vibrissae, notably some species of fish (e.g. catfish) who use them to help detect food and birds (e.g. swallows) to help detect objects in their environment.

Star-nosed mole

The star-nosed mole is specially adapted to living underground and its unusual nose also has a highly developed sense of touch. There are 22 appendages that surround its nostrils, which are covered in around 30,000 special touch receptors called Eimer's organs. These allow it to very quickly understand its environment and find prey. The star-nosed mole's somatosensory system is so complex that it is more similar to other mammals' visual systems.

Taste and smell

Jacobson's organ (also called the vomeronasal organ) is a specialised organ that forms part of the olfactory systems in many reptiles and mammals. It is located just above the hard palate, under the nasal cavity. It can detect specific chemicals called pheromones from prey, predators or mates.

Figure 56 The flehmen response in a horse

Figure 55 A star-nosed mole

When pheromones are detected, signals from Jacobson's organ are sent to the hypothalamus via a second olfactory bulb, called the accessory olfactory bulb. This is in contrast to the main olfactory system and olfactory bulb, which sends signals to the cerebral cortex. As discussed in section 3.1, the hypothalamus regulates hormones in the body, and thus chemicals detected by Jacobson's organ can have a direct and immediate impact on behaviour. By sensing pheromones, and then influencing behaviour, it plays an important role in the mating behaviour of animals.

Jacobson's organ has a complimentary role to the main olfactory system. This is because the main olfactory system is able to detect chemicals that easily evaporate into a gas - these are known as volatile compounds. Jacobson's organ, however, can analyse chemicals which do not easily evaporate into a gas, and remain in the form of a liquid - known as non-volatile compounds.

Mammals use the flehmen response to move air into their mouths. To do this they curl up their top lip, close their nostrils and lift up their heads. These actions expose the Jacobson's organ inside the mouth to pheromones.

Snakes use their vomeronasal organ to help them hunt by flicking their forked tongue out of their mouth to pick up chemicals. Inside their mouth they place the tongue onto the vomeronasal organ, which gives the snake

Figure 57 The pores of the ampullae of Lorenzini are clearly visible on the snout of this shark

information about their prey or a mate. As they get closer they can detect more pheromones, and the forked tongue gives informatuon about direction. This helps snakes hunt at night.

> **Did you know?**
> Humans also have a Jacobson's organ – however it is a vestigial organ, which means it no longer functions or plays any role.

Electroreception

Electroreceptors are a type of sensory receptor that detects the presence of electric fields. Electrical signals travel around the nervous systems of all animals, which generates a small electrical current in water. Electroreceptors detect this electric current, which is a signal that a living creature is nearby. Hence predators in water can use electroreceptors to detect the presence of prey.

Electroreceptors only evolved in aquatic animals. This is because water allows electrical currents to flow through it. Electrical currents cannot flow through air   any land animals equipped with electroreceptors would find them to be useless. As a result, land animals do not have electroreceptors.

Electroreception organs

Ampullae of Lorenzini

The **ampullae of Lorenzini** are an electroreception organ that is found in cartilaginous fish such as rays, skates and sharks. They consist of a series of pores that appear on the surface of the skin (Figure 57). Each pore is connected to a tube, filled with a gel that conducts electricity, which ends in a pouch called an ampulla. The ampulla is connected to nerves which send signals to the brain. (Cartilaginous fish have a skeleton made from cartilage rather than bone).

Just before sharks bite their prey, they roll their eyes back in their socket to protect themselves. Electrical signals from the prey's muscles are detected by the ampullae of Lorenzini around the shark's snout as they bite them. Sharks can detect tiny electrical currents from very far away. It is also believed they can detect the Earth's magnetic field, to help them with navigation and migration.

Platypus bill

The bill of the duck-billed platypus also houses tens of thousands of electroreceptors. It uses them to locate prey when foraging at the bottom of rivers and streams, because it closes its eyes, ears and nose when hunting. The duck-billed platypus is one of only five known mammals that can sense electric fields. (The others are the four species of echidna and the Guiana dolphin.)

> **Did you know?**
>
> Lorenzini was a 17th century scientist who first discovered these structures. 'Ampullae' is the plural of 'ampulla', which is the name of an ancient Roman jug. The shape of the pouches reminded Lorenzini of these jugs, which is where the name came from.

Echolocation

Another non-human sense is **echolocation**. It uses high-pitched sounds which humans cannot hear, called ultrasound. Animals that use echolocation send out a high-pitch sound. If it hits an object, it is reflected back, and the animal detects this echo. Ultrasound reflects differently from objects with different shapes, and allows animals using echolocation to hunt or navigate where vision is limited.

Dolphins, beluga whales and toothed whales make high-pitched clicks in the larynx which are amplifed in the melon, a fatty section in their head. If the ultrasound reflects off an object, the echo causes their lower jaw to vibrate. The jaw is connected to the animal's ears so the animal hears the echo, and can perceive the size, shape and speed of the object. This allows them to hunt in murky or deep water, where light barely penetrates.

Bats use a similar principle but they emit very loud high-pitched chirrups from their larynx instead. This ultrasound hits objects

Figure 59 An echo is a reflection of sound waves.

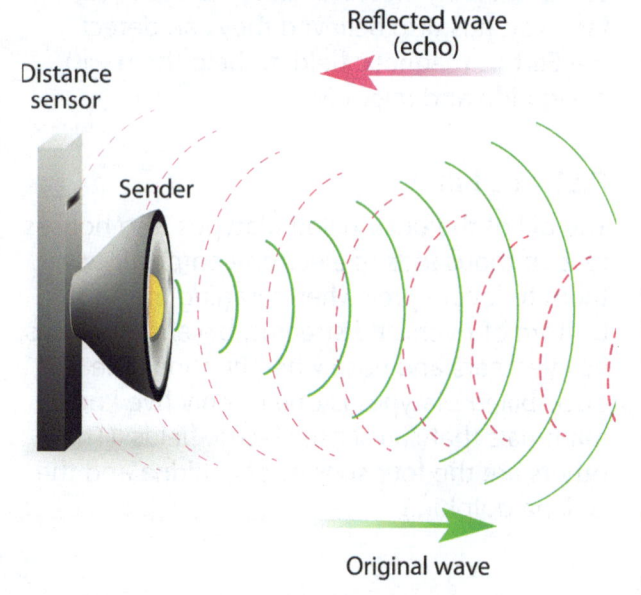

Figure 58 The grey long-eared bat has particularly large and distinctive ears. Its prey can hear ultrasound so its echolocation 'screams' are much quieter than other bats, to allow it to sneak up on prey - hence it needs very sensitive ears to hear the echo.

and reflects back. Bats' specially adapted ears (or nose in some species) allows them to detect tiny differences between the original ultrasound and the echo. This allows them to build up a detailed picture of their environment that is based on sound, not light. This means they can navigate and hunt at night.

As sound travels faster in water than air, dolphins can use echolocation to detect prey from further away than bats.

Other animals that use echolocation include

- oilbirds
- swiftlets
- common shrew.

However, their use of the system is less sophisticated than in bats and dolphins.

Infrasound

Infrasound is very low sound frequencies, below the range of human hearing. Elephants use infrasound to communicate over long distances, using low rumbles which they detect through the ground using their feet. Hippos and whales can also use infrasound.

Questions

1. a) State the name of a tactile organ. b) Describe how it functions. (2)
2. A bat goes hunting at night. Explain how it can successfully catch its prey. (4)
3. Describe the role that Jacobson's organ has in finding prey or a mate. (3)
4. How does a shark use different senses when hunting prey? (3)

Learning Outcome 4 – Revision checklist

4.1	☺	😐	☹
I can describe the structure, function and stimulus of the eye: the cornea, the pupil, the iris, the ciliary body, the lens, the sclera, the retina, rod and cone cells, the choroid, the fovea, the optic disc, the optic nerve, the medial and lateral rectus muscles.			
I can label a diagram of the eye.			
I can compare typical adaptations of eyes and eyesight between predatory and prey species.			
I can describe the structure, function and stimulus of the ear: the pinna, the auditory meatus, the tympanic membrane, the malleus, the incus, the stapes, the oval window, the round window, the cochlea, the organ of Corti, the cochlear nerve and the semi-circular canals.			
I can label a diagram of the ear.			
I can compare typical adaptations of ears and hearing between predatory and prey species.			
I can describe the structure, function and stimulus of the nose: the nasal chambers, the turbinates, the olfactory nerve, the olfactory bulb.			
I can label a diagram of the nose.			
I can discuss some adaptations of the nose and the sense of smell.			
I can describe the stimulus of the sense of taste and the structure and function of the mouth: the taste buds, the soft palate and the hard palate.			
I can label a diagram of the mouth.			
I can compare typical adaptations of the sense of taste between predatory and prey species.			
I can describe the structure, function and stimulus of the skin receptors in the sense of touch.			

4.2	☺	😐	☹
I can describe specialist tactile organs and why they have evolved, including: the platypus beak, lateral line and vibrissae.			
I can describe specialist organs for the sense of tase or smell, and why they have evolved, including Jacobson's organ.			
I can describe electroreception organs and why they have evolved, including ampullae of Lorenzini.			
I can describe the sense of echolocation and why it has evolved in bats and dolphins.			
I can explain why specialised senses or sense organs have evolved in two separate species living in contrasting environments.			

Learning Outcome 4 - My Revision Goals

Consider each of the points in the revision checklist. For each point, what do you still need to do to make sure you are fully prepared for your exam?

On a separate piece of paper, or using the space below, write down each of the tasks you identify. (If this book was given to you by your tutor then ask before writing in it!)

Glossary

adrenal glands
An endocrine organ associated with the 'fight or flight' response, and which releases three hormones: adrenaline, cortisol and aldosterone.

adrenaline
A hormone which causes the 'fight or flight' response. It acts on most tissues in the body, for instance causing the heart rate to increase, the pupils to dilate, breathing rate to increase etc.

adrenocorticotropic hormone
Stimulates the adrenal gland.

afferent system
The part of the peripheral nervous system that carries sensory information from the body towards the central nervous system.

air sacs
Structures within a bird's respiratory system that draw air in and push air out of the lungs.

aldosterone
This hormone acts on the kidneys to regulate the amount of salt in the bloodstream.

alveoli
Tiny sacs at the end of the bronchiole in the lungs. Alveoli are once cell thick to allow oxygen and carbon dioxide to pass through them and to/from blood capillaries.

ammonia
A toxic waste product produced by the body's cells.

ampullae of Lorenzini
An electroreception organ found in fish with skeletons made of cartilage rather than bone (such as rays, skates and sharks). It is seen as pores on the skin.

anoestrous
A non-reproductive period when the oestrous cycle stops altogether.

anti-diuretic hormone (ADH)
This regulates the amount of water in the blood. It acts on the kidneys and blood vessels.

aorta
A large artery through which oxygenated blood is pumped to the rest of the body.

appendicular skeleton
All of the bones which are not part of the axial skeleton. These bones append to (attach to) the axial skeleton.

arteries
Blood vessels which take blood away from the heart. They have thick walls.

atlas
First vertebra in the neck which allows the head to move up and down.

atrioventricular node
A component of the heart, located between the right atrium and right ventricle, which passes on electrical signals from the sino-atrial node into the ventricles.

auditory meatus
A passageway that connects the pinna to the middle ear.

autonomic nervous system
The part of the peripheral nervous system that is

responsible for bodily functions that are not under voluntary control – e.g. digestion, breathing and heart rate.

axial skeleton
The collective term for the skull, the vertebral column the sternum and the ribs.

axis
Second vertebra in the neck which allows the head to move side to side.

axon
A long fibre within a neuron that transmits electrical signals away from the nucleus.

axon terminal
The end of an axon in a neuron, that can form a synapse with the dendrites of other neurons.

B-cells
Another name for B-lymphocytes.

biconcave
In a red blood cell, when both sides of the cell curve inwards

bicuspid valve
Valve between left atrium and left ventricle in the heart. Also known as the 'mitral valve' or 'left AV valve'.

bladder
A sac in which urine collects until it is ready to be emptied.

blastocyst
A very early stage of development in an unborn animal. A blastocyst is a collection of cells that implants into the wall of the uterus, and which will develop into the embryo and placenta.

B-lymphocytes
Part of the adaptive immune system, they are a type of white blood cell which forms and matures in the bone marrow and gathers in lymph glands. Also known as B-cells, these can locate and destroy pathogens in the blood or lymph.

book lung
A respiratory organ in some invertebrates made up of a series of plates. It does not function like a lung.

Bowman's capsule
Part of the kidneys that receives filtrate from the glomerulus

bronchi
The two main passages that carry air into the lungs. They are made up of rings of cartilage and smooth muscle.

bronchioles
Branches within the lungs that form at the end of the bronchi.

bulbus glandus
A knot of erectile tissue at the base of a dog's penis.

bundles of His fibres
Components within the walls of the ventricles of the heart that transmit electrical signals that cause the heart to beat.

calcar
An extra bone on the foot of bats.

calcitonin
A hormone secreted by the thyroid gland which reduces the amount of calcium in the blood.

capillaries
Very small blood vessels that connect the arteries and the veins, and which allow oxygen and carbon dioxide to reach all the cells of the body. Their walls are only one cell thick, which allows oxygen and carbon dioxide to pass through.

Cardiac muscle
Non-voluntary muscle that only occurs in the heart.

carpals
Bones that make up the ankles/wrists in the foreleg.

central nervous system
The brain and the spinal cord.

cerebellum
Part of the hindbrain associated with balance and coordination.

cerebral cortex
A part of the forebrain responsible for higher-level processing. This includes problem-solving, retrieving memories, and interpreting sensory information.

cervix
The entrance to the uterus, this ring of muscle prevents infection entering the uterus.

chordae tendinae
Cords of tissue in the heart that are connected to the bicuspid and tricuspid valves.

choroid
A layer of tissue between the retina and sclera.

cilia
Small hair-like projections on the surface of cells that help trap pathogens.

ciliary body
A ring of tissue at the front of the eye which is connected to the lens and which contains a muscle to change the shape of the lens.

circulating hormones
Hormones which act at a distance from where they were secreted, travelling through the blood until they reach target organs or tissues.

clavicle
The collar bone.

cloaca
An orifice that birds have. It used to expel liquid and solid waste, as well as for reproductive functions.

closed circulatory system
A circulatory system where the fluid that circulates (e.g. blood) is completely enclosed within the vessels that make up the system, e.g. in mammals and fish.

cochlea
A spiral structure within the inner ear filled with a fluid. Movement of this fluid stimulates hairs in the organ of Corti, which is contained within the cochlea.

cochlear nerve
This nerve carries all of the electrical signals gathered from the organ of Corti and transports them to the brain for interpretation as sounds.

collecting duct
Final section of the kidney which transports urine to the ureters. ADH acts here.

cone cells
A light-sensitive cell that can distinguish between different colours.

copulation
The act of sexual intercourse or mating.

cornea
A thin, transparent disc at the front of the eye.

corpus luteum
The remains of the follicle after ovulation.

cortisol
The so-called stress hormone which gives the body's cells an energy boost by changing the way they process nutrients.

dendrite
A series of small extensions from the nucleus in a neuron which can form a synapse with axon terminals of another neuron.

dermis
The middle layer of the skin that contains blood vessels.

diaphragm
A muscle which facilitates the act of breathing. The diaphragm is unique to mammals.

dioestrous
The stage in the oestrous cycle when the corpus luteum breaks up.

distal tubule
Site of selective reabsorption in the kidney.

double circulatory system
A circulatory system where the fluid that circulates (e.g. blood) passes through the heart twice during a complete circuit, e.g. in mammals.

echolocation
The ability to send high-frequency sounds and detect their echoes in order to provide information about the surroundings. Used by bats and dolphins.

ectotherms
Animals that rely on their environment to maintain body temperature, e.g. reptiles.

efferent system
The part of the peripheral nervous system that carries motor information from the central nervous system towards muscles in the body.

embryo
The embryo is a collection of cells that will go on to become a foetus. The embryo stage is when all of the major body organs and systems are formed.

endocrine glands
The organs of the endocrine system that secrete hormones.

endocrine system
This describes the complete system of glands that secrete hormones in order to control organs elsewhere in the body.

endotherms
Animals that rely on internal metabolic processes to maintain body temperature, e.g. mammals.

epidermis
The top layer of the skin.

epididymis
A tube at the back of the testes that connects to the vas deferens

erythrocytes
Red blood cells.

excretory system
The body system which removes ammonia and its derivative substances from the body.

femur
Upper hindleg bone.

fertilisation
The stage in the reproductive cycle when a sperm and an ova (egg) fuse together to form a new cell called a zygote.

fibula
The thinner of the two lower hindleg bones.

Flat bones
A category of bones which are thin (though not necessarily flat). E.g. ribs.

foetus
A developing unborn animal. The foetal stage is after the embryo stage.

follicle stimulating hormone (FSH)
A hormone which stimulates the follicles to grow and mature. Secreted by the pituitary gland.

follicles
A structure within an ovary which contains a cell that can mature into an ovum (egg).

forebrain
Associated with higher level processes such as decision-making and emotional response. Contains the thalamus, the hypothalamus, the cerebral cortex and the limbic system.

fovea centralis
A small pit in the retina that only contains densely-packed cone cells. It is responsible for the sharpest image and the most detailed part of an animal's vision.

FSH
Follicle stimulating hormone.

gas exchange
This describes the process of oxygen passing through the walls of the alveoli and into the blood capillaries; and also the passing of carbon dioxide from the blood capillaries into the alveoli.

gestation
The period in the reproduction cycle where a collection of cells develops into an embryo and then a foetus, before it is eventually ready to be born.

gills
The main respiratory organ in fish, where gas exchange takes place.

glomerulus
A knot of blood capsules in the kidney that filters blood.

glucagon
A hormone which causes glycogen, a stored from of glucose, to be converted into glucose and released into the blood. It acts to increase blood sugar levels.

Graafian follicle
A developing follicle

haemolymph
A fluid used within the open circulatory system used by invertebrates. It performs a similar role to blood.

hair follicle
Structures in the dermis layer of the skin from which hair grows.

hard palate
The roof of the mouth towards the front, made up of bones.

hindbrain
Connects the spinal cord to the rest of the brain. It contains the medulla, cerebellum and pons.

homeostasis
The regulation of certain bodily conditions, so that they stay within acceptable limits.

hormones
Chemical messengers that are secreted by endocrine glands.

humerus
The bone of the upper foreleg.

hypodermis
The bottom layer of the skin that contains fat cells, also known as the subcutaneous layer.

hypothalamus
A small area of the brain that monitors a range of parameters such as body temperature, heart rate etc. The hypothalamus mainly secretes hormones that control the pituitary gland, although some act directly on other organs.

ilium
One of the bones of the pelvis.

implantation
The phase of the reproductive cycle when the blastocyst embeds into the uterus.

incus
The second of three bones in the ear, located between the malleus and stapes.

insulin
A hormone which causes glucose to be absorbed by the body, which decreases blood sugar levels. (Glucose is a type of sugar.)

iris
A ring of muscle in the eye which can widen and narrow to let more or less light through to the back of the eye.

irregular bones
A category of bones which do not fit into the other categories. E.g. pubis, ilium and ischium.

ischium
One of the bones of the pelvis.

islets of Langerhans
Regions in the pancreas that produce hormones. Consists of alpha cells, which produce glucagon, and beta cells that produce insulin.

Jacobson's organ
A sense organ that is part of the olfactory (smell) system in many animals. It used to detect the scent and pheromones of other animals.

joints
The points where bones meet.

keel
An extended ridge of bone on the sternum which is present in bats to help with flying.

kidneys
The main organs of the excretory system. They separate blood into its constituents, filter out anything unwanted and reabsorb substances back into the blood.

larynx
Organ at the top of the trachea responsible for the production of sounds. Also known as the voice box.

lateral line
A sense organ in fish which

can detect the movement of water.

lateral rectus muscles
Located in the eye and used to look outwards.

left atrium
Chamber of the heart into which oxygenated blood flows.

left ventricle
Chamber of the heart from which oxygenated blood is pumped to the aorta.

lens
A transparent and flexible tissue in the eye which can change shape in order to bend light, which focuses an image on the retina.

leukocytes
White blood cells.

LH
Luteinising hormone.

ligaments
Tissue that connects bones to each other.

limbic system
A collective term for the parts of the forebrain responsible for emotions and emotional responses.

local hormones
Hormones which are not released into the bloodstream and act on cells very near to where they were secreted.

long bones
A category of bones which are longer than they are wide. E.g. the humerus.

loop of Henle
Section of the kidney where water and salts are removed. It has an asending and descending limb.

lungs
The respiratory organs of most animals (though not fish) in which gas exchange takes place.

luteinising hormone (LH)
A hormone which helps one follicle to become dominant, completely mature and ultimately to ovulate, becoming a corpus luteum in the process. LH is secreted by the pituitary gland.

lymph
A colourless liquid that circulates through the lymphatic system.

lymph nodes
Also known as lymph glands, these organs connect lymph vessels and act as filters to trap foreign or unwanted bodies. Lymphocytes are located in the lymph glands. Swollen lymph glands are a sign of infection.

lymph vessels
Thin tubes containing lymph that, along with lymph glands, make up the lymphatic system.

lymphocytes
A special type of white blood cell responsible for the adaptive immune system. There are B-lymphocytes and T-lymphocytes.

malleus
The first of three small bones in the ear, connected to the tympanic membrane.

marsupials
Mammals who give birth to an undeveloped foetus. The foetus continues its development in a pouch. Marsupials have a less complex placenta. E.g. kangaroos, koalas.

medial rectus muscles
Located in the eye and used to look inwards.

medulla oblongata
Part of the hindbrain responsible for autonomic functions such as breathing and digestion.

meninges
Layers of connective tissue between the skull and brain.

metacarpals
Bones that connect the phalanges to the carpals in the forelimb.

metatarsals
Bones that connect the phalanges to the tarsals in the hindlimb.

metoestrous
The stage in the oestrous cycle when the corpus luteum is dominant.

midbrain
Connects the hindbrain and forebrain. Associated with the reticular formation.

monotremes
Mammals who lay eggs, whose embryos receive nutrients from the yolk in the eggs, not through a placenta. There are only three species of monotremes: duck-billed platypus, short-beaked echidna, long-beaked echidna.

nasal chamber
The interior of the nose, which is the entrance to the respiratory system. It warms and moistens air, and contains mucus and cilia to trap pathogens and prevent them entering the lungs.

negative feedback loops
A process which acts to negate any change to a system. They are used in all homeostatic process to ensure that parameters always stay within an acceptable range.

neuron
Also known as a nerve cell, this is one of the main cells that make up the nervous system. Neurons transmit and receive electrical and chemical signals. Different neurons have different specialities – e.g. motor neurons and sensory neurons.

neuroreceptors
Chemicals on the dendrites of neurons which bind to neurotransmitters.

neurotransmitter
Chemicals stored within synaptic vessicles which are released, travel across the synaptic gap and bind to neuroreceptors on a receiving neuron.

nodes of Ranvier
Gaps in the myelin sheath around an axon in a neuron.

noradrenaline
Hormone produced by the adrenal glands which constricts blood vessels as part of the 'fight or flight' response

nucleus
A section of a cell which coordinates cell activities and contains DNA. It is separated from the rest of the cell by a spherical membrane.

oestrogen
A female sex hormone which is secreted by follicles.

oestrous
The stage in the oestrous cycle when ovulation occurs.

oestrous cycle
The reproductive cycle of most animals. It differs from the menstrual cycle in a number of ways.

olfactory bulb
Part of the sense of smell, this organ is able to interpret signals from the olfactory nerve as different smells, but sends on the information to the cerebral cortex in the brain for further processing.

olfactory nerve
Located at the back of the nasal chamber, this nerve senses the presence of different molecules and passes electrical signals on to the olfactory bulb for interpretation as smells.

open circulatory system
A circulatory system where the fluid that circulates (e.g. haemolymph) is not enclosed within any vessels. E.g. invertebrates and arthropods.

optic disc
The area in the retina where the optic nerve is connected, which does not contain any rod or cone cells. It is the location of a 'blind spot'.

optic nerve
The nerve that sends visual information from rod and cone cells to the brain.

organ of Corti
This organ has hair-like projections which are stimulated by movements of the membrane in the cochlea in the inner ear. Stimulation of these hairs causes

electrical signals to be passed to the cochlear nerve.

ova
Eggs. The singular of ova is 'ovum'.

oval window
A membrane-covered opening which connects the stapes in the ear to the cochlea.

ovaries
Organs which are responsible for the production of ova (eggs).

oviducts
Tubes which connect the ovaries to the uterus. Also known as fallopian tubes.

oviparous
Animals whose embryos develop in eggs which hatch outside of the body, e.g. birds.

ovoviviparous
Animals whose embryos develop in eggs which hatch inside the body, so that mothers give birth to live animals, e.g. boa constrictors.

ovum
Egg. The plural of ovum is 'ova'.

oxytocin
A hormone which causes the uterus to contract, which is therefore essential for birth. It also causes muscles to push milk out of the mother's teats.

palate
The roof of the mouth, made up of the hard palate and soft palate.

pancreas
An endocrine organ that regulates the level of glucose in the blood by producing insulin and glucagon.

parabronchi
Organs within a bird's respiratory system where gas exchange takes place. They play a similar role to alveoli but unlike alveoli they are not dead-ends but instead open at both ends. This makes for a more efficient respiratory system.

parasympathetic nervous system
One of two efferent pathways in the autonomic nervous system, this is also known as the 'rest and digest' system. It acts to calm the body down, conserve energy and digest food, in response to the perception of safety. It acts in opposition to the sympathetic nervous system.

para-thyroid gland
An endocrine organ which increases the amount of calcium in the blood, by secreting parathyroid hormone.

parathyroid hormone
A hormone secreted by the para-thyroid gland which increases the amount of calcium in the blood.

parturition
The process at the end of gestation, where the animal expels offspring from the body. Also called birth.

pelvis
The bone that supports the hindlegs, made up of four bones: the sacrum, pubis, ilium and ischium. The last three make up the hip.

penis
Male sex organ that is used to inseminate females.

peripheral nervous system
The collection of nerves that connect the central nervous system to the rest of the body. Unlike the central nervous system it is not protected by bones and is therefore more easily damaged.

phalanges
The bones in the toes/fingers.

pharynx
Also known as the throat, the pharynx funnels air towards the larynx.

pinna
The visible portion of the outer ear.

pituitary gland
The 'master gland', the pituitary secretes hormones that control the other endocrine glands in the body.

placenta
In mammals this is the organ which supplies the embryo and foetus with all the nutrients it needs during the gestation stage of the reproduction cycle.

placental mammals
Mammals who give birth to live young. These animals provide nutrients to embryos and foetuses through the placenta. Almost all mammals are placental mammals. Also known as viviparous animals.

plasma
The liquid in which the components of blood are suspended.

platelet
A component of blood responsible for blood clots.

pons
Part of the hindbrain which acts as a bridge between the hindbrain and the rest of the brain.

positive feedback loop
A process which acts to reinforce any change to a system.

prepuce
A sheath covering the penis.

progesterone
A female sex hormone, secreted by the corpus luteum which, amongst other things, prepares the body for pregnancy.

prolactin
A hormone which causes milk to be produced in mammals.

pro-oestrous
The stage in the oestrous cycle when the follicles are growing. It is before the fertile period.

prostate gland
An organ which secretes a fluid to protect semen. The urethra passes through the prostate gland, which expels semen into the urethra during ejaculation.

proximal tubule
A segment of the kidneys conneted to Bowman's capsule

pubis
One of the bones of the pelvis.

pulmonary artery
The artery that carries blood from the heart to the lungs. It is the only artery that contains deoxygenated blood.

pulmonary veins
The vein that carries blood from the lungs to the heart. It is the only vein that contains oxygenated blood.

pupil
The opening in the centre of the eye which lets light through.

Purkinje fibres
Components within the walls of the heart that transmit electrical signals that cause the heartbeat to reach the ventricles.

radius
One of two lower foreleg bones.

reabsorption
The process by which the kidneys add substances back into the blood.

receptors
These are proteins on the surface of target organs. They respond only to specific hormones and, once activated, amend the behaviour of the organ in question. The endocrine system would not work without receptors.

reticular formation
A complex network of neurons in the brain running through and connecting the midbrain, the forebrain and hindbrain. Its role is to receive and transmit sensory and motor (movement) information between the brain and the body.

retina
The inner layer of the back of the eye, made up of light-sensitive cells (rods and cones).

ribs
A series of bones in the chest which, together with the

sternum, form a cage which protects vital organs such as the lungs and heart.

right atrium
Chamber of the heart into which deoxygenated blood flows.

right ventricle
Chamber of the heart from which deoxygenated blood is pumped into the pulmonary artery.

rod cells
A cell that is very sensitive to light but which cannot distinguish between different colours.

round window
A second membrane-covered opening in the cochlea within the inner ear. Its role is to allow the fluid within the cochlea to move.

sacrum
The section of the spine that is covered by the pelvis.

scapula
The shoulder bone, which supports the forelegs.

Schwann cell
A cell that forms a myelin sheath around an axon in a neuron.

sclera
The outer layer of the eye – the 'whites of the eye'.

sebaceous gland
Glands at the base of hair follicles which secrete sebum to lubricate skin and hair.

semi-circular canals
These three tubes are part of the ear but control balance.

sesamoid bones
A category of bones which resemble sesame seeds. E.g. the patella (kneecap).

short bones
A category of bones which are roughly cube-shaped. E.g. carpals and tarsals.

single circulatory system
A circulatory system where the fluid that circulates (e.g. blood) only passes through the heart once during a complete circuit. The heart only has two chambers, one atrium and one ventricle. E.g. in fish.

sino-atrial node
A component of the heart, located at the top right atrium, that is the source of the electrical signals that cause the heart to beat.

skeletal muscles
Voluntary muscles that are used to move the bones of the skeleton.

skin receptors
Nerve cells in the skin, also known as cutaneous receptors. There are three main types: mechanoreceptors, thermoreceptors and nociceptors.

skull
A collection of bones that protects the brain.

smooth muscle
Non-voluntary muscles. They are used in the stomach, the intestines, the uterus, the bladder.

soft palate
The roof of the mouth towards the back, made up of muscle and tissue.

spiracles
Openings on an invertebrate's body which are connected to trachea.

stapes
The third of three bones in the ear, connected to the oval window.

sternum
The bone which the ribs are attached to. Also known as the breastbone.

sympathetic nervous system
One of two efferent pathways in the autonomic nervous system, this is also known as the 'fight or flight' system. It acts to prepare the body for imminent physical exertion in response to some perceived danger. It acts in opposition to the parasympathetic nervous system.

synapse
A junction between neurons, where an axon terminal from one neuron connects to the dendrite of another neuron.

synaptic cleft
The tiny gap in a synapse across which neurotransmitters travel.

synaptic vessicle
A membrane within the axon terminal of a neuron which contains neurotransmitters that are released into a synaptic cleft.

T3 (Triiodothyronine)
A hormone secreted by the thyroid gland which increases cells' metabolism.

T4 (Thyroxine)
A hormone secreted by the thyroid gland which increases cells' metabolism.

tarsals
Bones that make up the hindleg ankles.

taste buds
Made up of receptor cells, these can detect the presence of certain molecules in food and send electrical signals to the brain for interpretation as different tastes.

T-cells
Another name for T-lymphocytes.

tendons
Tissue that connect bones and muscles.

testes
A male reproductive organ where sperm and testosterone is manufactured.

testosterone
A male sex hormone produced by the testes, responsible for male sexual characteristics and the production of sperm.

thalamus
A part of the forebrain that acts like a relay station, sending sensory information from the body to the cerebral cortex where it can be interpreted and acted upon.

thermoregulation
The methods animals use to keep their body temperature within an acceptable range.

thyroid gland
An endocrine gland responsible for regulating the body's metabolism, which is the speed of chemical reactions in the body. This affects how quickly bodily functions can be carried out. It secrets the hormones T3 and T4, and calcitonin.

thyroid-stimulating hormone
Stimulates the thyroid to produce T3, T4 and calcitonin.

tibia
The thicker of the two lower hindleg bones.

T-lymphocytes
Part of the adaptive immune system, they are a type of white blood cell which are formed and mature in the thymus and gather in lymph glands. Also known as T-cells, these can locate and destroy pathogens within the tissues of the body.

trachea
A tube which carries air into the lower respiratory system. It is made of C-shaped cartilage and smooth muscle.

trachea (invertebrates)
A network of tubes within invertebrates which transport air directly to cells in the body. Trachea are connected to spiracles.

tricuspid valve
Valve between the right atrium and right ventricle. Also known as the 'right AV valve'.

turbinates
Networks of bone, tissue and blood vessels in the nasal chamber, shaped like shells. They play a role in warming and regulating the moisture content of air that is breathed in.

tympanic membrane
Also known as the eardrum, vibrations of the air in the auditory meatus causes this membrane to vibrate.

ulna
One of two lower foreleg bones.

ultrafiltration
The process by which the kidneys remove substances from blood.

urea
A waste product that is produced after processing ammonia, which is a toxic waste product produced by the body's cells. Urea is one component of urine.

ureter
A tube which the kidneys pass urine through, and which connects the kidneys and bladder.

urethra
A tube connected to the bladder whose other end is an external opening. Animals pass urine out of the body through the urethra.

uric acid
A waste product that is produced after processing ammonia, which is a toxic waste product produced by the body's cells. Uric acid is less toxic than urea and is produced by birds instead of urea.

uterus
The organ where a fertilised ovum (egg) embeds and grows into a foetus. Also known as the womb.

vagina
A muscular tube which receives the penis during copulation and connects the uterus to the vulva.

vas deferens
A tube that transports sperm from the epididymis to the prostate gland

veins
Blood vessels which carry blood towards the heart. They have thin walls and contain valves.

vena cava
Large veins through which deoxygenated blood flows back into the right atrium of the heart.

ventilation
Another term for breathing.

vertebrae
The individual bones that make up the vertebral column.

vertebral column
The collective name for the series of bones that make up the neck, spine and tail, which are all made up of vertebrae.

vibrissae
Special hairs whose function is to provide tactile (touch) information. E.g. whiskers.

viviparous
Animals who give birth to live young and provide nutrients to embryos and foetuses through the placenta

vulva
The entrance to the vagina.

zygote
A sperm and ova (egg) that have fused together. A zygote is a single cell that has a full set of chromosomes. It will begin to divide, go on to form a blastocyst and ultimately a foetus.

Useful words

accessory olfactory bulb
A secondary olfactory bulb that is connected to Jacobson's organ.

adaptive immune system
The part of the immune system that learns to fight off new types of pathogen and can remember how to do so in the future; it is the basis of immunity.

antagonistic pairs
The name for pairs of muscles that work against each other to move body parts in opposite directions. For instance, the bicep and tricep form an antagonistic pair, which allows your forearm to flex up and down at the elbow.

antibody
Special proteins that attach to antigens to disable pathogens.

antigen
Molecules on the surface of a cell which are recognised by antibodies.

autonomic functions
Processes in the body which do not require voluntary control e.g. breathing.

basophils
White blood cells which release histamine that causes inflammation.

bone marrow
Spongy tissue found in the central cavity of some bones.

cartilaginous fish
Fish whose skeleton is made of cartilage not bone.

cetaceans
An order of aquatic mammals that includes whales and dolphins.

chromosomes
Very, very tightly coiled strands of DNA molecules, arranged into 'X' shapes. DNA contains a unique combination of instructions for making proteins in the body and contains an individual's genetic code. Each species has a certain number of chromosomes, which are contained in every cell within the body.

cursorial animals
Animals that have adapted for running.

cursors
Animals that have adapted for running.

cutaneous receptors
Nerve cells in the skin.

cytokines
Proteins secreted by cells, used to send signals to other cells.

dendritic cell
A type of monocyte that presents antigens to T-cells.

deoxygenated blood
Blood which is has very little oxygen dissolved in it.

digitigrades
Animals that walk on their toes but not on the soles of the feet.

embryonic diapause
When a blastocyst does not immediately implant into the uterus but remains in a state of dormancy. Once implanted normal pregnancy resumes.

enzymes
Substances that speed up chemical reactions.

eosinophils
White blood cells that can attack larger pathogens like parasitic worms, but also cause allergic reactions.

epiglottis
A flap in the larynx which stops food from entering the trachea.

exoskeletons
External skeleton.

fallopian tubes
Part of the female reproductive system that is the site of fertilisation. It allows the ova (eggs) to travel from the ovaries to the uterus.

flehmen response
A curling of the top lip, performed by animals to expose their Jacobson's organ to a scent or pheromone.

glabrous
Hairless.

glia
A type of cell that is found in the nervous system – although they do not carry electrical signals they are just as important as neurons.

glycogen
A form of glucose that can be stored in the liver, it acts as an energy store which the body can access at any time.

haemoglobin
A substance which transports oxygen in blood.

hippocampus
A part of the brain where memories are stored.

insemination
The action of depositing sperm into the female reproductive system.

intercostal muscles
The name for a series of muscles within the rib cage that are essential for breathing.

invertebrates
Animals without a vertebral column (backbone or spine).

lacteals
A lymphatic capillary (similar to a blood capillary) which absorbs fat in the intestine and transports it with lymph elsewhere in the body, until it can be absorbed in the blood.

left atrioventricular (AV) valve
Another name for the bicuspid valve.

lumen
The centre of veins and arteries, through which blood flows.

mechanoreceptors
Nerve cells in the skin that respond to gentle or normal touch.

melatonin
A hormone that is responsible for regulating sleep patterns.

metabolism
The rate of chemical reactions in the body, which determines the rate of all bodily functions.

mitochondria
An organelle within the cells of the body that uses oxygen and glucose to produce a molecule called ATP. ATP is a form of energy used by the cells of every known lifeform.

mitral valve
Another name for the bicuspid valve.

monocytes
Once activated these become macrophages or dendritic cells; both play a role in identifying pathogens to lymphocytes and clearing up dead cells.

myelin
Fatty protective coating around an axon in a neuron.

neural tube
The structure in an embryo that becomes the central nervous system (i.e. brain and spinal cord).

neuromasts
Hair-like projections within a fish's lateral line that respond to movements of water.

neutrophils
White blood cells that can consume and destroy pathogens

nictitating membrane
A 'third eyelid' used by a number of animals to clean their eye instead of blinking.

nociceptors
Nerve cells in the skin that send signals that are interpreted as pain, in response to extreme forces or temperatures.

non-volatile compounds
Chemicals that do not easily evaporate into a gas and instead stay in their liquid form.

olfactory system
The sense of smell.

osmosis
A process by which a liquid or gas moves through a permeable barrier because of a difference in concentration of some dissolved substance.

ossicles
The collective name for the three bones in the ear.

oxygenated blood
Blood which is has lots of oxygen dissolved in it.

pathogens
Small organisms that cause disease e.g. virus or bacteria

permeable
A permeable material allows liquid or gas to pass through it.

pheromone
A chemical produced by an animal which can affect the behaviour of another animal.

pineal gland
An endocrine gland located in the brain that produces melatonin, which is a hormone that is responsible for regulating sleep patterns.

right atrioventricular (AV) valve
Another name for the tricuspid valve.

sauropsids
A class of egg-laying vertebrates which include all modern birds and reptiles, and extinct reptiles such as dinosaurs.

secreted
The production and release of liquid by the cells of the body.

septum
Thin piece of cartilage and bone separating the nasal chambers. It is also the name for the tissue that separates the left and right side of the heart.

somatic nervous system
The part of the peripheral nervous system which is under voluntary control.

somatosensory system
The sense of touch.

sphincter
Circular muscles that can open and close bodily passages.

spinal cord
Part of the central nervous system which is made up of nerve tissue and which extends from the brain through the vertebral column.

spleen
The largest organ of the lymphatic system which acts as a blood filter, is a site for lymphocytes and which controls the number of red and white blood cells in the body.

synapsids
A class of animals who have a single opening in the skull behind each eye. The only remaining example of synapsids are mammals; ancient synapsids are the ancestors of modern mammals.

tactile
Meaning 'to touch'.

tapetum lucidum
A membrane behind the retina which reflects light back into the eye.

thermoreceptors
Nerve cells in the skin that respond to temperature.

thymus
An organ of the lymphatic system located in the chest where T-cells (T-lymphocytes) mature.

tonsils
An organ of the lymphatic system located at the back of the throat.

ultrasound
Sound that is too high for humans to hear.

ungulates
Hoofed animals.

vestigial
An anatomical feature inherited from ancestors which no longer has a function.

volatile compounds
Chemicals that can easily evaporate into a gas.

vomeronasal organ
Another name for Jacobson's organ.

Index

A

accessory olfactory bulb 88
adaptive immune system 55
ADH 48, 50
adrenal glands 48, 67
adrenaline 48, 67, 68
adrenocorticotropic hormone 47
afferent system 66
air sacs 16
aldosterone 48
alveoli 13, 14
ammonia 29
ampullae of Lorenzini 89
amygdala 64
anoestrous 23
anti-diuretic hormone 48
antibody 54, 55, 56
antigens 55
aorta 8
appendicular skeleton 35
arteries 10
atrioventricular node 9
auditory meatus 76
autonomic functions 62
autonomic nervous system 66
axial skeleton 34
axon 65
axon terminals 65

B

B-cells 54
B-lymphocytes 54
basophils 7
bats 15, 41, 90
bicuspid valve 8
binocular vision 74
bladder 30
blastocyst 24
blood clot 8
blood sugar 48, 50, 67
bone marrow 34, 54
book lung 17
bronchi 14
bronchioles 14
bulbus glandus 19
bundles of His fibres 10

C

calcaneal 40
calcitonin 49
capillaries 11
cardiac muscle 37
carnivores 82
carpals 34
cell receptors 47
central nervous system 65
cerebellum 62
cerebral cortex 63
cervix 20
cetaceans 15, 41
chewing 39
chordae tendinae 9
choroid 72
chromosomes 25
ciliary body 72
circulating hormones 51
clavicle 35
cochlea 77
cochlear nerve 78
colour vision 73
cone cells 72
copulation 24
cornea 71
corpus luteum 22
cortex 48
cortisol 48, 67
cursors 43
cutaneous receptors 82
cytokines 55
cytotoxic T-cells 55

D

deltoids 38
dendrites 65
dendritic cell 55
diaphragm 14, 40
dioestrous 23
diving mammals 15
dolphins 90
double circulation system 8–9
duck-billed platypus 27, 86, 89

E

ear 76
echidna 27
echolocation 89
ectotherms 57
efferent system 66
egg-laying mammal 27
electroreceptors 89
embryo 25
endocrine glands 47
endocrine system 47–51
endotherms 58
eosinophils 7
epididymis 20
Eustachian tube 77
excretory system 29–32
　aquatic mammals 31
　birds 31
　desert mammals 31
exocrine glands 51
eye 71
eyesight adaptations 73

F

femur 36
fertilisation 25
fibula 36
fight or flight 67
flat bones 36
flehmen response 88
foetus 25
follicle stimulating hormone 21–23, 47
forebrain 63
fovea 73
FSH 21–23, 47

G

gas exchange 14
gestation 25
gills 16
glabrous skin 83
glia 66
glucagon 48
glucose 48, 50, 68
gluteals 38
glycogen 48, 67
growth hormone 48
gustatory system 80

H

haemolymph 12
hard palate 39, 81
heart 9
heartbeat 9
helper T-cells 55
hindbrain 62
hippocampus 64
histamine 51
homeostasis 51, 58
hopping mammals 42
hormones 21, 47
humerus 34
hypothalamus 21, 47, 50, 58, 59, 64, 88

I

ilium 36
immune system 54
implantation 25
incus 77
inner ear 77
insulin 48
intercostal muscles 14
iris 72
irregular bones 37
ischium 35

J

Jacobson's organ 88
joints 37

K

kidneys 29, 50

L

lacteals 53
larynx 13
lateral line 86
lateral rectus muscles 73
latissimus dorsi 38
left atrioventricular valve 8
left ventricle 8
lens 72
LH 22, 47
limbic system 63
local hormones 51
long bones 36
lungs 14
luteinising hormone 22, 47
lymph 52, 54
lymph glands 52, 54
lymph nodes 52
lymph vessels 52
lymphatic system 52–56
lymphocytes 7, 54

M

malleus 76
mammalian respiratory system 13–15
marsupials 26
masseter 38
mechanoreceptors 82
medial rectus muscles 73
medulla oblongata 62

medulla (adrenal gland) 48
melatonin 23
metabolism 48
metacarpals 35
metatarsals 36
metoestrous 23
midbrain 63
middle ear 39, 76
mitochondria 37
mitral valve 8
monocytes 7, 55
monotreme 26, 86
motor neurons 66
mouth 80
musculoskeletal adaptations 39–41
musculoskeletal system 34–43
  disease 40

N

nasal chambers 79
neuromasts 86
neuron 65
neuroreceptors 66
neurotransmitters 66
neutrophils 7
nictitating membrane 74
nociceptors 82
nocturnal animals 74
non-volatile compounds 88
nose 79
nucleus 65

O

oestrogen 21, 22, 49
oestrous 23
oestrous cycle 21
olfactory bulb 79
olfactory nerve 80
olfactory system 79
omnivores 81
open circulatory system 12
optic disc 73
optic nerve 73
organ of Corti 78
osmosis 32
ossicles 77
osteoarthritis 40
osteomalacia 40
outer ear 76
oval window 77
ovaries 20, 49
oviducts 20
oviparous 26
ovoviviparous 26
ovum 21
oxytocin 26, 48

P

panoramic vision 74
para-thyroid gland 49
parasympathetic nervous system 67
parathyroid hormone 49
parturition 25
pathogens 52
pelvis 36
penis 19
peripheral nervous system 66
phalanges 35
pheromones 88
pineal gland 23
pinna 76, 78
pituitary gland 47, 50, 64
placenta 25
plasma 7
plasma cells 55
platelets 7
pons 63
prepuce 19
pro-oestrous 23
progesterone 21, 23, 25, 49
prolactin 25, 48
prostate gland 20
pubis 35
pulmonary artery 9
pulmonary veins 9
pupil 72
Purkinje fibres 10

R

radius 35
reabsorption 29
red blood cells 6
reproductive system
  cats 21
  female 19
  male 19
  pigs 21
  whales 21
respiratory system
  amphibian 15 16
  birds 16
  fish 16
  invertebrates 17
  mammals 13–15
rest and digest 67
reticular formation 63
retina 72
ribs 34
right atrioventricular valve 9
right atrium 8
right ventricle 9
rod cells 72
root hair plexus 82
round window 77
running mammals 43

S

sacrum 36
sauropsids 38–39
scapula 35
sclera 72
semen 25
semi-circular canal 78
sense of smell 79
sense of taste 80
sense of touch 82
sensory neurons 66
septum 79
sesamoid bones 36
sexual reproduction cycle 24
short bones 36
single circulatory system 11
sino-atrial node 9
skeletal muscle 37
skin receptors 82
skull 34
smooth muscle 37
soft palate 80
somatic nervous system 66
somatosensory system 82
sperm 25, 49
spinal cord 34, 65
spiracles 17
spleen 53
stapes 77
star-nosed mole 88
sternum 34
sympathetic nervous system 67
synapsids 38–39
synaptic cleft 66
synaptic vessicles 66

T

T-cell 54
T-lymphocyte 54
T3 49
T4 49
tactile organs 86–90
tactile system 82
tapetum lucidum 73
tarsals 36
taste buds 80, 81
testes 20, 49
testosterone 20, 49
thalamus 63, 64
thermoreceptors 82
thymus 53, 54
thyroid gland 49
thyroid-stimulating hormone 47
tibia 36
tonsils 53
trachea 13
  (invertebrates) 17
tricuspid valve 9
turbinates 80
tympanic membrane 76

U

ulna 35
ultrafiltration 29
ultrasound 89
ultraviolet light 74
ungulates 43
urea 29
ureter 30
urethra 19, 29
t 29
uterus 20

V

vagina 20
vas deferens 20
veins 10
vena cava 8
ventilation 14
vertebrae 34
vertebral column 34
vibrissae 87
volatile compounds 88
vomeronasal organ 88
vulva 20

W

whiskers 87
white blood cells 6

Z

zygote 24

101